CRYPTOCURRENCY INVESTING BASICS

An Introduction to Blockchain, Digital Assets, and Risk Awareness

AF808152

Arkon Flux

TABLE OF CONTENTS

ISBN: 9781807651404

Published by TPSA Publishing

WHY YOU SHOULD READ THIS BOOK

Most people receive new ideas with great fear, scepticism, and even denial. Of course, it is much safer to stay in the comfort zone and keep walking the beaten track. However, at some point, maybe it is worth asking yourself: how far will this track lead me? Most likely, it will lead you no further than the next door of your life.

Movement is progress. Therefore, you should keep an open mind and learn new things in order not to decay as the world keeps moving forward. This is true for many areas of our lives and in personal finances even more so.

For one, paper money is no longer prevalent. You will probably agree it is more convenient to pay for our costs and comforts with the small plastic card issued by your bank. This type of cashless settlement is much better for security reasons as well.

Not so long ago, humanity discovered another, completely new type of money – digital money or cryptocurrency. This is a new generation of currency that is created through the use of encryption software. These units of cryptocurrency are formed and preserved through algorithmic encryption.

I would like to focus a little bit on the kinds of problems that cryptocurrency can solve. This is an issue I will bring up throughout this book, but for now I want to focus your attention on a concept called *trust*. It is this important problem which cryptocurrency solves. Let me explain using an example from my life.

I decided that I wanted to create a business and make some money when I was in school. Three school friends and I set out to sell flowers. Each of us wrote down the information about his earnings in an old notebook, which we passed on to each other. When this notebook ended up in my hands, and I really wanted ice cream □, I wondered if I could change the written figures for my own benefit. But I scolded myself and thought: what if my friend did the same? Thus, it is obvious that distrust can arise even between the closest of friends. Cryptocurrency solves this problem as it does not allow us to add or change something once it is in the system.

However, the emergence of cryptocurrency, which solves the huge problem of trust, should not be associated with the explosion of a comet, a gift from above, or any other sort of fairytale beginning. In fact, the creation of cryptocurrency is much simpler.

First, the corresponding technologies, such as EXAMPLE, appeared, and then cryptocurrency arose on their basis, followed by mining (the production of cryptocurrency). Only afterwards did tokens, ICOs, and various cryptocurrency business models came into being on the basis of this whole ecosystem. The registry, which protects against manipulation (i.e. database repository), became the very first technology in the cryptocurrency economy. There is one small (or rather big) peculiarity of such registries: you can enter any data into this database but cannot forge anything or enter backdated information. These self-enforcing contracts are the second important technology in the cryptocurrency world. They are *tightly sewn* to the database.

Right from the start, I want to warn you that the cryptocurrency market is alive and changing. It works 24 hours a day, seven days a week. Therefore, the information which was up-to-date the day I wrote the book may not be quite as fresh the day you read it. I will recommend resources in this book that you may refer to in order to keep up with the times.

The last thing to mention here: I suppose there are sceptics among the readers of this book. I agree that we should meet all new ideas while maintaining a degree of scepticism, but we should also be reasonable. Therefore, before making hasty conclusions about cryptocurrency, I suggest carefully reading the book and then analyzing the information I share here. With that being said, put your verdict on cryptocurrency into cold storage for now.

If you are ready to plunge into the world of cryptocurrency, come on!

CHAPTER 1. MYTHS ABOUT CRYPTOCURRENCY AND THE MAIN RULES OF THE CRYPTOCURRENCY MARKET

Before setting goals in the cryptocurrency business, first let's go through some common misconceptions around cryptocurrency.

Cryptocurrency Myths

I frankly admit that I also almost got infected with scepticism towards cryptocurrency at the beginning of my journey, as many people scared me by telling me that cryptocurrency is a wildcat venture, financial pyramid, or the like. Such false beliefs often fence many people off from possible success. Only a small number of sticklers verify such statements about cryptocurrency instead of just blindly believing them.

I am sure that you have also heard allegations that no country will ever recognize cryptocurrency so cryptocurrency has no future. To refute this, it is enough to recall the recent increase in the value of cryptocurrency. Some countries have already recognized cryptocurrency as means of payment. Thus, this so-called "entertainment for geeks" has evolved into very real business for investment bankers and various corporations.

Therefore, if you, personally, still do not earn money in cryptocurrency, it is most likely because of the large number of the aforementioned statements, which you may have heard from pseudo-experts.

Let's enumerate the statements that stop many people halfway:

1. It's too risky. What if I invest money and cryptocurrency is banned worldwide the next day?
2. I do not have any financial education so I wouldn't know what I was doing;
3. My friend/brother/neighbor invested money in cryptocurrency and lost everything;

4. The topic of cryptocurrency is too new and difficult to understand as it is so I'd better wait a couple of years when everything becomes clear.

All these statements are merely excuses for your inaction. While you come up with additional excuses, people all over the world have already built a business in the cryptocurrency market and are making a profit. Ask yourself: what makes me worse than these people? If you realize that you are not worse than your friend/brother/neighbor, who already earns money in the cryptocurrency market, then ask yourself the following question: why should I enter the world of cryptocurrency business right now?

Why You Should Enter The Cryptocurrency Business World Now

So I will answer this question point by point. Each of the arguments will begin with the words **right now** so that you better understand why you should do this **right now**.

First, **right now** there is very little competition in the market. Are you surprised? It's true. Yes, the interest in this topic constantly grows, but in general, the competition is still very low. This market is still quite "wild" and unmastered. The reason for this is that most people tend not to trust new trends and ventures, as we have already mentioned above.

Second, **right now** you can reap high returns on this market. Cryptocurrency prices are at the stage of growth and development, but even so, you can earn a decent sum of money. As long as you manage your risks wisely, you can make profits in time.

Third, **right now** there are a lot of freeloaders in the market. They come to the cryptocurrency market to make a quick buck without delving into the subject. Usually, such people come from the pyramid schemes led by so-called "experts" who teach people without having any real experience in the topic themselves. I must admit that the market encouraged such people in the beginning, when they had a chance to make lots of money then. However, nowadays, if you want to achieve something like this in the cryptocurrency market, you will have to make real efforts.

Fourth, **right now** there are low risks in the market. Five years ago, most people waited for the day when cryptocurrency would be officially banned. These days, many countries around the world have already recognized the relevance of cryptocurrency. Cryptocurrency has gained such momentum that no one can simply press the stop button now. Do you really think that this fact does not prove you can invest your money without fear of various prohibitions and restrictions?

Finally, before moving on to the rules of the game in the cryptocurrency market, I suggest considering the future of cryptocurrency through the example of the Tesla company. Currently, the cost of a Tesla is estimated in the amount that, according to the expert forecasts, it will not be able to make back its expense in 300 years. Why do smart experts rate Teslas so highly then? Let's puzzle it out.

These days, an electric car is beautiful and stylish but expensive and not a very practical vehicle. However, experts are not concerned about today. They visualize a future that can become a reality in 20 or 30 years. It is difficult to imagine in years to come the ordinary cars we see daily in the streets now. You may imagine your old beloved car in the future, but personally, I picture some kind of semi-flying vehicles with solar panels or something even more sophisticated. Therefore, it is the electric vehicle that will have a bright and successful future, and Tesla is likely to occupy a leading position in the market.

It turns out that, by and large, no one knows for certain whether Tesla will hold its position in the market for the next 20-30 years. However, many people strongly believe it will, and this belief makes them invest in the company. Therefore, personally, I can easily believe that Tesla will hold a monopoly in the entire automotive manufacturing sector in 20 years.

How is this example linked to the future of cryptocurrency? Nowadays, some people view cryptocurrency (like an electric vehicle) as meaningless – a fashionable, interesting and technically curious trend. But a trend, nonetheless. However, this tiny drop now is likely to grow into a large dominant sea in the future.

Setting Goals And Identifying The Main Rules Of The Market

Before breaking new ground, real professionals learn the rules of the game and set goals, determining what kind of result they want to achieve. Since we are professionals, we will first deal with this task.

The following points must surely be among the cryptocurrency trading goals you will set for yourself:

- define the specific sum of money or percent of your income you will invest monthly in the area of cryptocurrency;
- define the degree of your readiness for risks;
- define your specific goals in the short and long-term.

Having set the goals, it is important to understand what is needed for a successful and fast startup in the cryptocurrency market. You will be surprised, but the theory is the least important thing in this field. Many people say they lack the information and knowledge for doing this type of business. However, most of the strategies I will tell you about do not require deep knowledge of the cryptocurrency business world. It will be enough for you to have a grip on basic principles of the cryptocurrency economy.

Then what's the problem? You need practice. Only practice, not a book (not even mine), will help you to understand where to buy, where to sell, how to store, and how to transfer cryptocurrency.

You will also need:

- Several ready-made strategies with low risk to enter the market
- Ability to filter content and information around you
- Communication with more experienced traders of the cryptocurrency market and "spying" on their actions
- Risk management
- An audit from an experienced curator.

And now we proceed to the statement you must remember once and for all: making any investment in the cryptocurrency market IS A RISK. If you are not ready to accept this, don't bother trying. Any opinion or forecast for the development of a particular coin, the reliability of ICO (this will be discussed later) is only a biased stance. There is not a single person across the globe who could give you an iron-clad guarantee for further developments. There are absolutely no right decisions and 100% guarantees. I or someone else can only give you a piece of advice, not guarantees.

You are the only person to bear responsibility for each decision. You should not subsequently blame an online resource, where you read about the prospects of some currency, or a friend who recommended you a reliable ICO, or even me for your possible financial losses! I, as the author of this book, will share my thoughts about reliable cryptocurrencies with you. However, this will be just my subjective opinion, again, only at this very moment. Therefore, if you are up to the task of blaming others for your possible failures, not yourself, you'd better close this book immediately and waste no time.

The cryptocurrency market is alive and constantly changing. Therefore, when dealing with cryptocurrency trading, one must learn to take personal responsibility for making decisions and always remember that no gains can be made without taking risks. Yes, there are less risky strategies, but risks exist in any case. Invest just the sum you are willing to lose without much regret.

And right now, lay the book aside and write down the two rules you should never break:

1. Do not invest to the last penny.
2. Have a stash of cash to take advantage of opportunities.

And finally, beginners should learn several safety points:

- When trading on the exchange, protect your account with two-factor authentication and keep your code name private.
- Your password should have at least 26 characters; a special password generator may help you come up with one.
- Never keep all your money on one exchange or in one wallet.
- Trade only on tried-and-true exchanges.
- Do not use public access points to trade in the cryptocurrency market.

CHAPTER 2. BASIC PRINCIPLES OF CRYPTOCURRENCY AND THE BATMAN OF OUR TIMES

Let's take a roundabout approach to this topic, starting from the banking sector.

The entire banking system of our days, regardless of the country, is arranged in such a way that we do not own our money. Central banks in any state hold emission monopoly, which is provided through legislation, creating a sure stumbling block to cryptocurrency legalization.

Fiat money is a term used by the cryptocurrency community to designate a currency without intrinsic value as money by government regulation or law (dollar, euro, etc.).

In theory, this currency should be secured by at least those goods, products, or services that are produced in the territory of a particular country (GDP) so that every citizen of this country can change their personal money for products. All the banks of the country also undertake the same currency, and the central bank promises to maintain its stability and reliability. This is how it works in theory, but no one can actually guarantee the currency's stability.

The government of a country is the main customer of all goods and services for the population, *i.e.* one of the largest employers. It is also the largest customer for the construction of roads, houses, hospitals, schools, *etc.* This makes up the lion's share of the GDP of each particular country. Accordingly, the government ensures the life of the population – pays pensions and social benefits. All these funds are taken from the central bank, which can issue currency and finance the government.

A policy dubbed *quantitative easing* was once implemented in the USA and later even spread to Europe and Japan. Due to this phenomenon, the amount of money in the world has greatly increased, while the purchasing power of the dollar has decreased over the past 100 years by 95%. This trend is ongoing. The more money exists, the cheaper it becomes.

It is important to keep in mind that each kind of currency experiences inflation, which indicates the depreciation of currency over some time. In other words, inflation is the speed of money circulation. The country's economy has its own cycles as people take on loans and then pay them back.

I cannot help but mention these loans. In the USA, the loan interest rates are at an average of 1%. They remained at zero level for a long time in the past. The central banks issued money and bought up financial assets, and accordingly, the poor people remained poor, and those people who had financial assets, stocks, or real estate made a constant profit. That is why each crisis makes the poor even poorer and the rich even richer. These processes result in a very powerful stratification of the population. However, alas, the economy works in this way, and we approach a kind of dead end, as many financial analysts state. All of this to demonstrate that the central banks do not cope very well with their function.

So people started to look for an alternative way to preserve the value of money. If not to multiply, at least not to lose. That's why many people invest in gold, fixed income instruments, shares, *etc.* At the same time, the economy began to develop swiftly after the Internet appeared, so people realized they do not need to hold money in physical form now. Thus, the concept of electronic money emerged.

The idea of creating a digital currency like Bitcoin is not entirely new. Still, there is a difference between Bitcoin and other types of digital money. If you use such electronic money systems as PayPal, Western Union, or Skrill, your finances are stored in the same companies. In this case, we deal with centralized money management, *i.e.* the fate of your money depends on decisions of specific people in these companies. You have no power to influence these decisions.

The Advantages of Cryptocurrency

Cryptocurrency is quite a different thing. This decentralized currency is characterized by the independence from a single transaction processing center. It is very difficult to track cryptocurrency transactions and impossible to cancel them. Using this type of currency, two people can carry out a purchase and sale transaction directly on the Internet, without resorting to the center of financial transactions.

But let's discuss in more detail the advantages of cryptocurrency over fiat money. These advantages are obvious!

☐ **Emission and circulation standards**: cryptocurrency is established once, and it is inviolable, while the standards for fiat money are changed arbitrarily by the central banks.

☐ **The issue** is carried out: cryptocurrency flows from the network to the participant while fiat money flows from the central bank to banks, from banks to companies, and only from companies to participants.

☐ **The flow of funds**: direct for cryptocurrency; through banks, payment systems, and cash for fiat money.

☐ **The number of participants**: 5 million for cryptocurrency; 7 billion for fiat money.

☐ **Transaction speed**: high for cryptocurrency; low for fiat money.

☐ **Anonymity**: always possible for cryptocurrency; sometimes possible when dealing with cash for fiat money.

☐ **Inflation**: impossible only for cryptocurrency; a constant reality for fiat money.

☐ **Rate volatility**: definitely high for cryptocurrency; low for fiat money.

The Inventor of Bitcoin And Why He's the Batman of Our Times

In the midst of the global financial crisis of 2008, somebody under the name of Satoshi Nakamoto designed Bitcoin and created its original reference implementation. He released the first Bitcoin software that launched the network and the first units of the Bitcoin cryptocurrency. Bitcoin became a new type of digital currency, which was very different from all the others. Its main difference lies in the fact that it is decentralized. Therefore, each participant cannot influence its fate.

The following question arises. There are a lot of people on the network, they do not know each other, and it's logical that they do not trust each other. So how can they be sure that payments are carried out and their money will not be stolen? However, everything was thought out in advance by Satoshi Nakamoto and carefully described by me in the book *Mastering Bitcoin for Starters*. Suffice it to say that Nakamoto's vision for a decentralized currency held true, solving many problems.

By the way, the identity of Satoshi Nakamoto is still shrouded in mystery. Several attempts have been made to disclose it because some people believe that Satoshi Nakamoto is a group of people rather than one person. However, none of these attempts has proved successful.

CHAPTER 3. BITCOIN AND MINING

To puzzle out the essence of Bitcoin, we need to plunge a little bit into the subject of mining. I will tell you about mining as an investment strategy later on, but for now let's consider mining only in terms of the emergence of Bitcoin.

A couple of years ago, many people viewed mining, the producing of cryptocurrency or Bitcoins, as a kind of hobby. Many played that *game* until the famous situation with pizza happened. That day is known as the Pizza Day in the cryptocurrency community. On May 22, 2010, a developer paid a fellow Bitcoin forum user 10,000 BTC for two pizzas. Back then, Bitcoin was almost worthless. These two pizzas cost the user approximately $25. Currently (October 2017), one Bitcoin costs more than $5,000. You can easily calculate that those two pizzas cost that guy millions of dollars.

Bitcoin was first mined on an industrial scale when its value began to zoom up. First, computing power was used to mine Bitcoin. Later, miners started to use graphics cards to form the blocks. Now, they use specialised equipment called ASIC. The largest such farms are now located in China.

How the Mining Process Works

Mining is the process of producing new cryptocurrency or bitcoins. Let's sort out how the mining works. When, for example, you send money from your wallet to another person's wallet, your transactions fall into the so-called "mempool." The Bitcoin mempool is a collection of all transactions waiting to receive a network confirmation. Miners, guided by their own principles, collect transactions in specific blocks and then try to insert them in the Blockchain

A miner gets a reward for mining a block. Currently, the reward is 12 bitcoins. That is, if you introduce a well-formed block into the system, you will get such a reward. By the way, the very first (Genesis) block was created by Satoshi Nakamoto. Any person could find it and get 50 bitcoins for it. Today, the reward for a properly formed block begins to fall. It is said to tumble to an infinitesimal value by 2140.

One day, miners realized that it was not profitable to mine on a standalone basis. The likelihood you will find such a block depends on "hash rate" or how powerful your Bitcoin miner's machine is. Under certain acceptable values, if say, your hash rate makes up 10% of the total hash rate, you will be able to find such blocks with a probability of 10% and get your reward. So, if you mine just using your laptop at home, you'll never find a new block. Therefore, in order to get a more stable reward, the miners unite in the so-called mining pools. They use their united hash rate to reap more regular profits.

Now many miners began to think: Why should I mine if I can buy Bitcoin? It is also a good idea as the value of Bitcoin is determined only by the belief of those people who use it, while its cost is determined by demand. If no one buys your Bitcoin from you, it will cost nothing. Therefore, as long as people see this technology as an opportunity to use it anonymously, make large payments, and so on, the value of Bitcoin will gain momentum.

For now, I will consider the advantages of Bitcoin, which still raise some questions.

The first advantage is **low transaction costs**. However, we should bear in mind that Bitcoin is not good enough for micropayments. If you transfer someone $1 million, it will cost you a penny. However, if you decide to pay for a cup of coffee, the fees will be large compared to the cost of your coffee.

The second advantage is **high-speed transaction processing**. Here are also some hitches. In fact, every block in the Blockchain generates every ten minutes, *i.e.* the fastest possible transaction takes ten minutes. It seems this is rather quick if compared to the SWIFT transfer in a bank, which can take 2-4 days. However, Visa and MasterCard are much faster. They can process tens of thousands of times more transactions per unit of time than Bitcoin. It is worth noting that some other types of cryptocurrency now exist which were developed to be faster than Bitcoin.

The third advantage is the **pseudo-anonymity of participants**. We have already studied that anyone can track all transactions on the network. If you know the exact owner of a wallet, you can track absolutely all transactions carried out from it. Thus it cannot be said that Bitcoin is absolutely anonymous. As soon as one manages to match your address with your personality, anonymity disappears. But if you observe so-called "internet hygiene," meaning you do not show your wallet to anyone, then basically your transactions cannot be tracked. I should mention, though, there are other cryptocurrencies which are more anonymous than Bitcoin.

By the way, if you want to stay anonymous when making transactions on the Bitcoin network, this is already possible. There are many services that wash your bitcoins. They are called "mixing services" and are used to mix one's funds with other people's bitcoins, intending to confuse the trail back to the funds' original source. As you see, new technologies have appeared to step up the anonymity of cryptocurrencies.

And, finally, now more than 16 million bitcoins circulate in the world, while a total of 21 million will be mined. Such figures are programmed by the algorithm of the network itself. The limited amount of Bitcoin makes inflation of this currency impossible. This currency does not depreciate over time because only a certain amount will be issued. Bitcoin even has a deflationary model: many people lose their coins by forgetting the password for the wallet or sending money to the wrong address. Therefore, the number of bitcoins will gradually decrease.

Proof Of Work abd Why It's Important

The biggest problems of the Bitcoin network were as follows: how to make sure that the transactions are really truthful; how to make sure that a miner does not deceive anyone; what should be done to choose the right block and actually build the Blockchain. All these issues are settled via the *consensus algorithm*.

The **Proof of Work protocol** confirms that a miner does a huge amount of work to find a correct nonce and get a successful hash. You should spend much time to find the needed one. I'll explain in more detail.

The block difficulty is adjusted every 2016 blocks and depends on how many zeros are at the beginning of a particular hash. It is not difficult to find the hash itself, but it takes pains to find a successful hash with a certain number of zeros. If you have the hash of a previous block as well as timestamp and transaction data, it seems that it is very easy to make a new hash from this and process this block. However, you need to find a nonce, whose value is set so that the hash of the block will contain a run of leading zeros. Much time is needed for that. Once miners find this successful hash, they send a block to the Blockchain. That is, they have already confirmed all transactions, having done some work. So, there is no point in deceiving someone as such work is very difficult to do.

After that, all information is distributed in the nodes. First, a miner sends one node. It can run a check on whether, for example, those people who sent money from point A to point B really had that money, *i.e.* whether all the transactions are valid. Then the nodes start to exchange this information with each other, and thus, the block is formed.

In theory, it could happen that two miners create one and the same block. How will Blockchain choose which block is better? The first principle is speed. The second principle is "success" of a hash. Therefore, "success" of a hash is exactly the efforts which the miners should make within the Proof of Work protocol.

Another reason why you need to choose a "successful" hash is an adjustment of the network difficulty. The more miners appear, the more network difficulty grows, which means that transactions can be processed more quickly. If the miners slow down in finding the blocks, the difficulty goes down.

Let me add a few more words about how to settle the situation when several miners create identical blocks on the network. The essence of the Blockchain consensus is that the longest chain of blocks is considered to be fair. If the blocks begin to be built in a direction different from yours, then your first block will again fall into the pool of unconfirmed transactions. This often happens when the network is overloaded. So to make sure that the next blocks are built exactly under your block and you will get the reward, wait until several more blocks are formed after yours. If more than five blocks are formed, the money is definitely yours.

At the end of the previous subsection we discussed the pros of Bitcoin, so now it's time to talk about the cons of Blockchain, namely the 51% attack or double-spending. I will explain.

Any mining pool can unite its efforts to such an extent that the probability of generating the next block in this pool can be 51%. The cryptocurrency community once witnessed the situation when the members of the Chinese mining pool artificially restricted new members in their system and managed to generate about six blocks in a row. It was after this incident that it became clear you have to wait for the confirmation of transaction for one hour, not for ten minutes. That is if you create five blocks, and each of them is formed for ten minutes, we multiply by five and get 50 minutes.

CHAPTER 4. BLOCKCHAIN

Before we get into any of the technical details behind Blockchain technology, it's important to understand the problems that Blockchain solves. Why do we need Blockchain, and what does it do that our current technology can't do?

The early adopters of Bitcoin and Blockchain technology spotted what they perceived as a fundamental flaw in the way we think about transactions, trust, and social institutions. The earliest versions of Blockchain came right around the same time as the 2007 financial crisis in the United States, when many people lost faith in societal institutions that were supposed to protect the interests of the common man. Of course, people were disillusioned by the banking system in the wake of the crisis, but they also lost faith in the government to regulate financial markets and in the press to investigate potential crises.

Most people would agree that our institutions have flaws and aren't perfect solutions. But they do solve problems of trust, and they've been doing it for hundreds of years. In fact, we're probably living in the most peaceful, comfortable era in human history. Any alternative to our current institutions needs to have clear advantages and strengths.

The idea behind Blockchain is to replace institutions run by imperfect human beings with technology that can do the job better and also empowers individuals. If you could create a way for strangers to trust one another without needing a bank or a government as an intermediary, you'd tackle one of society's biggest bottlenecks. But in order to do so, you'd need a powerful system for creating consensus between strangers, and the creators of Blockchain believe that power lies in decentralization.

Basically all applications of Blockchain (and other cryptographic technologies) are based around the concept of decentralization. Instead of a rigid, slow central authority making decisions and governing relationships, Blockchain seeks to return regulatory power back to the individuals. Instead of trusting a major institution, Blockchain builds trust through consensus.

How Does the Blockchain Work?

In simplest terms, Blockchain uses a combination of cryptography and a public ledger to create trust between parties while maintaining privacy. Understanding the mechanics of how this works is a little bit more difficult, but in order to fully appreciate the genius behind Blockchain technology, we'll need to dive into the technical details.

While Blockchain can include many more features, the fundamentals of a Blockchain are in the technology's name:

The block: a block is a list of transactions from a certain time period. It contains all the information processed on the network within the past few minutes. The network only creates one block at a time.

The chain: each block is linked to the block before it using cryptographic algorithms. These algorithms are difficult for computers to calculate and often take several minutes for the fastest computers in the world to solve. Once solved, the cryptographic chain locks the block into place, making it difficult to change. We'll look at this in greater depth in just a minute.

The chain grows longer over time. Once a new block is created, the computers on the network work together to verify the transactions in the block and secure that block's place in the chain.

The most fundamental part of the Blockchain is the ledger. It's where information about the accounts on the network is stored. The ledger inside the Blockchain is what replaces the ledger at a bank or other institution. For a cryptocurrency, this ledger usually consists of account numbers, transactions, and balances. When you submit a transaction to the Blockchain, you're adding information to the ledger about where currency is coming from and going to.

A Blockchain ledger is distributed across the network. Every node on the network keeps its own copy of the ledger and updates it when someone submits a new transaction. This "shared ledger" is how Blockchain intends to replace banks and other institutions. Instead of having the bank keep one official copy of the ledger, everyone will keep their own copy of the ledger and then we'll verify transactions by consensus.

Each Blockchain technology has its own ledger, and the various ledgers work very differently (as we'll see). However the Bitcoin ledger, the first Blockchain ledger, requires three pieces of information to list a transaction:

1. An input: if John wants to send David a Bitcoin, he needs to tell the network where he got that Bitcoin in the first place. Maybe John received the Bitcoin yesterday from Sarah, so the first part of the ledger entry says so.
2. An amount: this is how much John wants to send to David.
3. An output: this is David's Bitcoin address and where the Bitcoin should be deposited

Now comes the concept that's difficult to grasp: there is no such thing as a Bitcoin. Of course, there are no physical Bitcoins. You probably already knew that. However, there are also no Bitcoins on a hard drive somewhere. You can't point to a physical object, digital file, or piece of code and say, "this is a Bitcoin."

Instead, the entire Bitcoin network is only a series of transaction records. Every transaction in the history of Bitcoin lives in the Bitcoin Blockchain's distributed ledger. If you want to prove that you have 20 Bitcoins, the only way you can do it is by pointing to the transactions where you received those 20 Bitcoins.

Almost all Blockchain have this characteristic in common. The transaction history is the currency. There's no difference between the two. Some new cryptocurrencies are altering the way the ledger is written in order to provide greater anonymity and privacy in transactions. They use certain identity masking techniques to hide the sender and receiver of the transaction while still maintaining a functional distributed ledger.

Creating a Block

The ledger is the core of the block, but it's not the only thing that goes into a newly created block. There is a header and a footer required for every block. Additionally, the transactions included in the block are put through a process that compresses, encodes, and standardizes them. When a verifier creates a new block, it looks completely different from the ledger it was based off of. However, the underlying ledger is still there and can be accessed in the future when new transactions require information about the previous blocks.

Adding Transactions

The first step in building a block is gathering and adding all the current transactions to the block's ledger. When a user creates a new transaction, they broadcast that transaction to the entire network. Then a verifier's computer will review the transaction to make sure it's valid.

Since Blockchain currencies are nothing more than a series of transactions, the first step to verify a transaction is to look at where the sender says they originally got their money. The verifier will then review the history of the Blockchain to find the block and transaction where the sender received the money. If that input transaction is confirmed on the Blockchain, then the transaction is valid, and they will need to confirm the receiving party's address. If the input transaction hasn't been confirmed, then the current transaction is invalid, and it won't be included in the ledger.

Once all the transactions in that block have been verified, it's time to create the ledger. Here's a simple example, where the transactions are listed one right after another:

[Input][Amount][Output address], [Input][Amount][Output address], [Input][Amount][Output address], [Input][Amount][Output address], [Input][Amount][Output address]...

Then the verifier will apply a cryptographic technique called hashing to each of the transactions. At its most basic definition, hashing takes a string of characters and generates another string of characters. So, when you feed the input, amount, and output address to a hashing algorithm, it will turn the transaction into a string of characters unique to that transaction, like this:

aba128d3931e54ce63a69d8c2c1c705ea9f39ca950df13655d92db662515eacf

(This is an actual transaction hash from the Bitcoin Blockchain.)

So hashing is used to standardize data while making sure that it hasn't been tampered with. If someone were to try to change a transaction in the Blockchain, they'd have to rehash that transaction, and it would look entirely different. It would be obvious that it had been tampered with.

To make it even more difficult to tamper with the blockchain and reduce memory required to store the transaction ledger, most blockchains hash more than once. This means that they take the hash of a transaction, combine it with a hash of another transaction, and re-hash that into a new smaller hash. Combining transactions in this way is known as a Merkle Tree, and the root hash of all the transactions is included at the beginning of the block. Understanding why we need a Merkle Tree is a topic for a more in-depth book, but on a basic level, the Merkle Tree shows that all the transactions in the block are valid while using less memory in the long run.

Time Stamp & Block ID

The final element in a block is the time stamp and any block ID information. This makes it easy to look up previous blocks later on. Future transactions will also be able to point to this block ID as the block containing the input transaction (also known as the "coinbase") for the current transaction.

Linking Blocks Together

The final step of creating a block is linking it to the previous blocks in the chain. There are a few ways to do this, but virtually all of them involve hashing in some way to make the content of the previous block part of the new block.

Remember that hashing takes an input, no matter how big or small, and turning it into a string of characters. If you change the input even slightly, the entire output gets changed. In order to include the previous block's contents in the new block, we can take the hash of the entire previous block and add that to the beginning of the next block. Doing so means that we've effectively linked the old block to the new block, because if anything changes in the older block, even the tiniest change, the entire block's hash will change.

Now, once a block has been completed, it becomes MUCH harder to change it. Making an edit to an older block means you'd have to re-hash that entire block. Once you re-hash all of block 1, you'd have to crack open block 2, delete block 1's old hash, insert block 1's new hash, and now re-hash all of block 2. But new blocks are being created all the time, so in order to change an older transaction, you'd have to edit every block after that transaction took place. The more time that goes by, the harder it becomes to hack the network and successfully change a transaction. This is why hashing is at the core of Blockchain security. The cryptography makes the transaction ledger difficult to change, meaning the ledger can be public and secure at the same time.

However, the hashing itself is not that difficult. Most computers could easily re-hash a Blockchain in a few seconds. So in order to guarantee the hashing security does its job, we need to introduce a level of difficulty to the creation of a new block. Ideally, it would be something that slows an attacker down and makes it more likely that honest members of the network will win. In the Bitcoin Blockchain (and most other modern blockchains), that added difficulty is called "proof of work."

CHAPTER 5. WALLETS OR HOW TO SECURELY STORE BITCOIN

People dealing with cryptocurrency use a wallet as a safe depository and an instrument for incoming and outgoing payments. Let's analyze the available types of wallets and choose the most suitable one based on your computer's resources and tasks.

There are hot and cold wallets. There are also warm wallets, but they are used much less often. Cold wallets are used to store money, while hot wallets are used to send and receive the currency quickly.

As a rule, a wallet has a Private key and a Public key. The Private key belongs only to you, and you should never show it to anyone. You must keep it in mind as you sign all transactions with this key. At the same time, someone can use public keys to transfer money to your account, for example, for a new Ferrari car □. In such a case, you should give this person your public key. This key can even be published on social networks. There is nothing to worry about here.

Bitcoin Core will be the very first wallet to consider. It is the original Bitcoin wallet from the legendary Bitcoin creator Satoshi Nakamoto. It is the only officially supported wallet which is constantly updated by the professional Bitcoin community and stores the entire Bitcoin database on your computer, automatically supporting the network. Bitcoin Core is highly secure and user-friendly. However, I do not recommend installing this wallet on your computer. The first synchronization takes a very long time, and the size of a fully synchronized wallet reaches 100 GB, which is a significant drawback.

I consider Blockchain to be the ideal wallet for users, especially lazy ones. It does not require the installation of third-party applications on the computer. It allows you to create a Bitcoin wallet within seconds and use it immediately. This wallet guarantees a high level of safety for your Bitcoin, enjoys an impeccable reputation, and offers 24/7 support. The interface is simple and intuitive, even for beginners. The theoretical disadvantage lies only in the fact that your Bitcoin wallet is located on a third-party resource, not on your computer.

MyEtherWallet is another kind of good wallet. It is the most popular wallet for participation in ICOs (Initial Coin Offering). We'll go through ICOs later in the book, but for now, it's only important to know that it's very similar to an IPO.

It is simple enough in use: we choose the address of a person whom we want to transfer money to, indicate the sum, and transfer it. This wallet is called "warm" because it contains a file you need to download to your computer and only then enter a login for your wallet.

We have examined only several types of wallets for Bitcoin and Ethereum, but there are quite a lot of them for other cryptocurrencies.

I would like to mention that you can use several wallets simultaneously: for storing money, making transactions, for one-time use only.

For example, you can store Ethereum and all ICO tokens related to it on the MyEtherWallet wallet. I also recommend using the Mist wallet when dealing with Ethereum, though it takes much time to download it to a computer. There are also multi-currency wallets that support several cryptocurrencies such as Metamask and Jaxx.

Cold Storage and Bitcoin Debit Cards

Assuming that you want to learn how to store cryptocurrency correctly, I will dwell in more detail on the issue of cold storage.

Cold storage is a term referring to private keys, which are usually created and stored in a secure, isolated environment. This means that your money will not be located on any site. This wallet looks like a USB flash drive: you can connect it to a computer and make a transaction quickly.

The most popular cold storage wallets are Trezor, KeepKey. and Ledger.

This method is very popular for storing a significant amount of funds in cryptocurrency. Transactions are seldom performed, and security is one of the top priorities. Therefore, the cold storage suits you the best if you decide to hold your funds for the long haul.

There is another reliable option allowing you to store your cryptocurrency – Bitcoin debit cards.

The essence of these cards is also quite simple. You register a wallet on the site and transfer bitcoins to it. After that, the service sends your Bitcoin debit card, linked to your mobile phone number, to the indicated address. It is very convenient to pay for purchases using this card. However, I warn you that this method of storage is not cheap. Each time you pay with this card, there's a fee of about 3%. Therefore, this method of storing Bitcoin is not entirely suitable for daily use. However, this card is the perfect choice for people who get their salary paid in Bitcoin and want to pay for their purchases with this card anywhere in the world.

Finally, remember that a wallet can be hacked, so any hot storage option carries a certain level of risk. Take the same approach as with your bank cards: do not store all your money on a single bank card. You need hot wallets to add money to your account on the cryptocurrency exchange and to make transactions. They are also used to participate in the ICO. However, I recommend you hold most of your money in a cold wallet. It is not connected to the Internet so it cannot be hacked, which makes it way more secure.

To make your wallet safe, create a separate email account for each wallet, protect it with two-factor authentication, and, most importantly, write down all your passwords on paper.

CHAPTER 6. ETHEREUM PLATFORM

Creation and How It Functions

One day, a Canadian programmer of Russian descent, Vitalik Buterin, reflected on the imperfections of Bitcoin. He realized that using Bitcoin (and cryptocurrency in general) for payment was only a small portion of its full potential..

Realizing the advantage of Blockchain, which is not subject to change, he came up with a new concept – **smart contracts**. Bearing in mind that it is possible to store absolutely any information in such a network, Buterin decided to create his own blockchain called Ethereum.

Ethereum's blockchain is similar to the Bitcoin blockchain except it can be used for things other than just transactions.

One of the main advantages of Ethereum's blockchain is that each new block generates in the network within only 15 seconds, not 10 minutes as in the case of Bitcoin. However, in my opinion, another important advantage of Ethereum's blockchain is that its creator is a real, live person. Vitalik Buterin is invested in the future of his platform, visits various cryptocurrency conferences, shares ideas, creates bank consortia, *etc.*

SMART CONTRACTS

Smart contracts can be created in Ethereum and solve a very important problem: how to make a deal on the Internet when you do not trust another party. How does it work?

A smart contract is a computerized transaction protocol that executes the terms of a contract. It often happens that one of the signatories may interpret the contract terms the way he likes it. Meanwhile, smart contracts are computer protocols intended to facilitate, verify, or enforce the negotiation or performance of a contract in an unbiased manner.

Thus, smart contracts allow for the regulation of people's relations on the Internet by:

- Providing computer-controlled monitoring and execution of agreements
- Reducing execution costs
- Contributing to the creation of more trustworthy relationships
- Doing without a third party to perform the terms of the contract
- Having a mathematical formulation and clear logic of execution
- Requiring an environment that allows one to fully automate the execution.

So let's think about where you can apply smart contracts. First, you can use them to buy or sell shares in the stock exchange and participate in crowdfunding. Nowadays, many companies have to pay more than $5 million for their shares to be traded on NYSE or NASDAQ. These costs are incommensurably smaller if using Blockchain, and they can theoretically be reduced even to zero.

It is also possible to hold a vote with the help of smart contracts to avoid ballot stuffing, re-voting, and any errors in vote counting. Furthermore, the results can be seen immediately.

The next thing you can do with smart contracts is to run an auction. Such auctions will be fully transparent, and any person will be able to take part. No kickbacks or corruption will be possible.

Smart contracts can even provide copyright and protect products against counterfeiters.

You can also use smart contracts to keep various registries. For example, the government of Sweden experimented with Blockchain to keep records of all land plots, as well as their purchase and sale, back in 2016.

Smart contracts can also do good for affiliate programs, insurance, gambling and lottery, transparent taxation, and much more. Almost all habitual interactions with government agencies can take advantage of the benefits of smart contracts using Ethereum's Blockchain.

As you can see, smart contracts offer a lot of opportunities, so this technology has every chance to get even greater use in the future.

However, smart contracts do have some drawbacks, namely, the problem of scaling. Bitcoin's Blockchain is limited. However, there are no restrictions in Ethereum.

And finally, lots of people worry about the PoS Ethereum. For those out of step, I can quickly explain that PoS Ethereum resembles Proof of Work in Bitcoin.

The transition from PoW to PoS is a significant milestone in the development of the Ethereum network., and it is inadvisable from an economic point of view to stay in PoW as it leads to high inflation over time.

Ethereum creator Vitalik Buterin has recently announced his intention to conduct a survey on Ethereum's transition to partial Proof of work The Ethereum's mining rewards will be halved, but new opportunities to verify the transactions will appear. A node that verifies transactions will have large quantities of Ether (cryptocurrency). That is, a person must have a certain amount of Ether available to be able to process transactions. Such a person is supposed to verify transactions absolutely honestly as he/she is also a holder of Ether.

As you might understand already, PoS is a sort of central bank which holds a large amount of currency. Therefore, Ethereum will partially turn from decentralization of Ether to an oligopoly. There will be large "holders" who have accumulated huge quantities of Ether and thus obtained the possibility to process transactions. This will be their way of making a profit.

Full transition to PoS is expected in early 2018. Since the power of the network will immediately fall and the reward for the block will drop, I predict that some members of cryptocurrency community will switch to another currency or even leave the market. I predict only the most powerful Ether miners will stay on the Ethereum's Blockchain.

CHAPTER 7. THE TOP 10 BEST CRYPTOCURRENCIES AND WHERE TO TRACK THEIR PRICES

Consider the important factors that influence the price of cryptocurrency.

Obviously, the first factor is the real benefit of using a particular currency. The second one is the belief in the growth of an ecosystem of this coin. The third factor is the belief in the growth of the cryptocurrency ecosystem as a whole, *i.e.* the belief that most businesses will want to use the cryptocurrency system. The last factors of pricing are the extra charges for the deficit and future profit as well as the risk discount.

Also, it happens quite often that various speculations and demand for a coin raise its price in disproportion to its intrinsic value. The most important thing here is to find out the percentage ratio between a soap bubble and something really efficient. There is no exact algorithm for calculating this ratio. Therefore, everyone solves this task relying only on own views.

I will give an example. If you want to find out what the total worth of McDonald's is, you simply enter this query into the Google search engine and any resource will give you the same information. However, it won't fly with cryptocurrency. The cryptocurrency exchanges are local, and cryptocurrency trading resembles the Forex market. In other words, it is an over-the-counter market. There is no central market where you can track the value of a certain currency. As a rule, members of the cryptocurrency community view prices at different sites, which vary in regional terms, and the price gap may reach ten to twenty percent.

Cryptocurrency is also open-source, so everyone can make sure of how they work on the GIT HUB site, the most popular resource that unites all open-source developers.

To choose cryptocurrency, I recommend using cryptocompare.com. It is an aggregator showing cryptocurrency prices on different cryptocurrency exchanges. Here you may read about each currency and view its current rate to another currency, including the US dollar.

Another useful site is coinmarketcap.com. It is convenient for watching the latest data on the top 10 best cryptocurrencies. But remember that this top ten is constantly changing.

The name of the currency can be seen in the first column (Name). The second column (Market Cap) shows capitalization of the currency. How is it calculated? It is very simple. The amount of currency that is circulating in the market now (Circulating Supply) is multiplied by the price (Price).

The Best Cryptocurrencies

If you want to invest in cryptocurrency, I recommend that you choose among top 10 or at least top 20 best cryptocurrencies. The rest are not reliable and can demonstrate both ups and downs. These currencies with lower rank have bigger volatility, posing a significant risk.

Volatility is good for professional traders since they are able to manage risks and the price dispersion offers greater opportunities for making money. At the same time, volatility is bad for a novice or an investor. That is why most funds tend to buy only the top 10 best cryptocurrencies. Yes, sometimes new cryptocurrencies, which have only recently appeared in the market and already managed to attract a huge amount of money, hit this list. But they do not stay among best for long. An example of such a cryptocurrency is IOTA. The capitalization of this currency has already declined almost twofold since it entered the market.

I would also like to say a few words specifically to Bitcoin fans. Now there is a huge amount of different cryptocurrencies in the market, so it would be very foolish to fixate on just one option. Indeed, Bitcoin is the most reliable cryptocurrency and is growing rapidly. However, these days there are plenty other interesting cryptocurrencies apart from Bitcoin. Each of them has its own advantages.

If you're not sure what kind of cryptocurrency you should buy, you can order an audit of its code.

After all, as I wrote earlier, all cryptocurrencies are open-source: they utilize the open-source code and are similar to each other. In fact, any person can get the code for Bitcoin, for instance, change some parameters, and thus invent a new cryptocurrency. To be blunt, this is the way the fork appears

The fork is the use of the code base of a software project as a start for another one. Each such project can develop independently from the main one and can realize the opportunities which the main project lacks. The process of creation of such cryptocurrencies is called forking.

Bitcoin naturally serves as a basis in the world of cryptocurrency, so all other coins are called Bitcoin forks.

Almost all new cryptocurrencies are copies of already known cryptocurrencies. Therefore, it turns out that it is actually impossible to have a diversified investment portfolio, *i.e.* to distribute risks by holding more than one currency.

Now we finally review some of the best cryptocurrencies today.

Litecoin is the first currency I want to focus on. In my opinion, it is one of the safest currencies after Bitcoin. It was created in 2011 and is a Bitcoin fork, *i.e.* the code of this coin is based on the Bitcoin source code. Litecoin block generates every 2.5 minutes. About 84 million Litecoins are now in circulation, which is four times more than the number of Bitcoins. In addition, Litecoin already has SegWit, which was activated on the Bitcoin network only on August 1, 2017.

Lighting Network has also been recently activated in Litecoin. Lighting Network is an off-chain micropayment system with exceptionally low fees. If you think it is not important, then just believe me, it is really of a landmark nature.

In short, Litecoin is a modern, fast-paced currency. I recommend you have both Bitcoin and Litecoin in your investment portfolio. Bitcoin as a very reliable currency and Litecoin is used for conducting experiments which can later be applied to Bitcoin.

Zcash is another quite interesting coin to pay attention to. It is the most anonymous coin available today. It uses zero-knowledge proofs, a building block for greater financial and data privacy in cryptocurrencies. Only professional developers understand how it works. Nevertheless, the protocol of this coin is now used in many Blockchains.

The value of this currency was immediately more than $4,000 when it appeared. In other words, the recoupment of mining for one day was 100%. However, the price of this coin fell sharply after a while.

Here I want to move your attention to a very important issue. People are often inclined to buy up cryptocurrency, falling for the hype. In other words, they buy a particular coin only because a lot of other people do so, not realizing that can be done for speculative reasons, not because the coin has any real value. As a consequence, you can be in the red for a very long time. This is why I recommend you not fall for the hype of a new kind of cryptocurrency and just keep your coins in your wallet so you can earn much more.

Now let’s proceed to **Ethereum Classic** currency. Let's figure out how it actually appeared.

In June 2016, perhaps the most massive attack in the history of the cryptocurrency industry happened. A very promising and popular project The

DAO lost more than $60 million due to a mistake in the code. But let's back up a moment first.

The DAO is the largest crowdfunding project ever. It was the first Ethereum-based decentralied autonomous organization and a form of investor-directed venture capital funds. When buying tokens in this organization, you got into a certain company, which you could remotely control by voting. Put simply; it was a whole corporation where people could reach agreements, choose investment instruments, and then get dividends from the company.

However, one day a certain person found a loophole in The DAO algorithm and stole tokens worth about $60 million. Later, this person stated he had done nothing illegal as he had not cracked the code. He even warned he was ready to defend his position in court.

The action of this person provoked a real panic because many users lost their money. Also, the situation with The DAO very much affected the Ethereum itself. The Ether price immediately dropped by 50%. That's why founder of Ethereum Vitalik Buterin was forced to make a centralized decision to save the situation. He decided to create a fork.

There are two types of fork: soft fork and hard fork. Soft fork is a change to the software protocol where only previously valid blocks/transactions are made invalid. Since old nodes will recognize the new blocks as valid, a soft fork is backwards-compatible. Hard fork is a radical change to the protocol that makes previously invalid blocks/transactions valid (or vice-versa). Hard fork is a permanent divergence from the previous version of the Blockchain, and the newest version will no longer accept nodes running previous versions.

Vitalik Buterin suggested creating a soft fork, *i.e.* to roll back all transactions on the network until the moment of the theft. Initially, all The DAO members consented to his proposal, but then some people appeared who said they were not ready for such a centralized change. It is because of those people the hard fork took place. Ether split into two currencies. The first and basic is Ethereum Classic.

By the way, there are some forecasts that after Ethereum switches from Proof of Work to Proof of Stake, as we mentioned above (i.e. when part of the coins will be mined, and the nodes will guarantee the other), many miners will earn less on Ether. They are likely to switch to Ethereum Classic. I think this situation can cause the growth of this currency.

Now let's consider **Dash** currency. This currency is a tool for intraday trading. This coin is very technical regarding charts. It works on a partial Proof of Work and Proof of Stake system.

At the same time, it is a wonderful currency for everyday spending. Also, Dash offers a masternode tool, which is only expected to be launched on Ethereum platform.

This currency has a great future if it will be the first one to provide an opportunity to transfer money easily and pay for purchases. For example, a project called Dash Evolution is being developed now. It is habitual online banking, which is already called the best alternative to PayPal. It is expected to be completed by the end of 2017. I predict that the project will be praised by users who do not support cryptocurrency anarchism or rigid decentralization but just want to take advantage of cryptocurrency for making transactions and holding their money.

Another great advantage of this currency is anonymity. The coins are mixed with every new transfer of this currency, making it rather difficult to track a transaction on your wallets. Bitcoin has a separate service for mixing coins, but this happens automatically when using Dash.

The next coin for discussion is **Waves**, created by Russian programmer Alexander Ivanov. The Waves platform allows you to create your own tokens in a few clicks and sell them immediately. They are traded on the decentraliszd Waves exchange, and any person who buys tokens from you can immediately try to sell them on this exchange. It will take just 5-10 minutes to create your own tokens on Waves.

If we compare Waves tokens to Ethereum tokens, the first ones are simpler since they do not require complex programming.

Another interesting coin is **Dogecoin**. It is a meme coin, which was invented in jest. No one expected it would circulate. In any case, this coin has no value and is practically never used.

The Dogecoin price is deliberately screwed up now and then, and people gladly participate in this amusement. The person who is the last to buy the coin at the highest price, loses money.

And a few words about **Namecoin**. It is a Bitcoin fork which allows you to buy a .bit website, *i.e.* a decentralized hosting.

Let me also mention some rather unreliable currencies. **Ripple** is an example of such cryptocurrency (although many have already earned much on it).

Ripple is a centralized currency. It is controlled by a consortium of banks, the largest of which are located in Japan. This coin started to grow after Bitcoin had been legalized in Japan.

A significant disadvantage of this currency is that you lose a certain amount of Ripple when, for example, you transfer dollars or euros to someone. And so the question arises: how does this affect the exchange rate, depending on the number of transactions?

At the same time, the official website of this currency offers no wallets. If you want to create a wallet, you will have to look for it on gatehub.net.

Now let's summarize. At the time of writing this book, a total of 804 cryptocurrencies exist. But I do not advise you to study all of them, just the top ten. You can invest beyond thistop ten only when you start making profit.

Let me give you some practical advice. So before choosing a currency for investment, I recommend reflecting on the composition of your investment portfolio (I'm talking about passive portfolio investment). If you want to have the top three best cryptocurrencies in it, then consider what can affect the price of these currencies shortly. You also need to carefully analyze the markets on which this currency is traded and watch the daily charts to estimate the turnover of your currency in last 24 hours.

Finally, I want to remind you once again that cryptocurrency market provides no guarantees. If you have read information about more or less reliable coins, it does not mean this information is true. This is just my subjective opinion. Therefore, it is up to you to decide whether to buy particular coins or not.

CHAPTER 8. THE MOST POPULAR CRYPTOCURRENCY EXCHANGES

Now that we've sorted through the features of different coins, we can move on to learning about the most popular cryptocurrency exchanges.

You need to use the the exchange if you want to buy or sell cryptocurrency for dollars, euros, or other conventional currency as well as exchange one cryptocurrency for another.

I urge you to work only with the most popular exchanges. You can be more confident in their reliability, which means you will keep your money safe.

Poloniex is the first exchange worth paying attention to.

Poloniex is one of the largest American exchanges for Bitcoin trading. According to the coinmarketcap.com aggregator service, it ranks second in transactions turnover among all world cryptocurrency exchanges trading Bitcoin. Despite this fact, Poloniex is among the exchanges that often malfunction (from what I've seen). Some people suspect it may soon go bankrupt.

Let's move on to some practical tips about trading on exchanges.

It is impossible to deposit conventional money onto the Poloniex exchange, *i.e.* dollars, euros, and so on. You can buy tokens to exchange on tether.to. These tokens are practically not subject to change. Such stability makes me think that people will prefer holding their cryptocurrency assets precisely in these tokens in case of trouble. It would be a kind of safe haven.

Anyway, the easiest way to enter this exchange is to deposit cryptocurrency. After you have registered on the exchange, you will have a very limited functionality available consisting of the amount of money you can deposit and withdraw from the exchange.

By the way, you will need to undergo verification. Verification is necessary to limit the possibility of drug traffickers, arms dealers, or money launderers using the exchange. Therefore, the people, who want to trade in = great volume are invited to undergo verification. It looks like this.

Verification takes place in several stages, depending on the degree of desired anonymity. For example, four levels of verification will allow you to withdraw about $400,000 per month. But there is one setback. It may take a few weeks for your application to be reviewed considering the huge volume of users on the exchange It will take another couple of weeks to undergo verification and deposit money. Moreover, there is no guarantee your application will be approved.

This exchange is a bit authoritarian for my taste. If its owners see problems with cryptocurrency on their exchange, they simply prohibit withdrawing it. People who own this currency have to bend over backwards to sell and withdraw it in another cryptocurrency.

Here are some other most popular exchanges to date.

Kraken is a large European cryptocurrency exchange with a huge trade volume. I consider Kraken to be the fastest and most user-friendly platform.

Bitfinex is a popular Hong Kong cryptocurrency exchange. It offers a few more tools than other exchanges and has a more sophisticated design.

Bittrex differs from others as it is the so-called altcoin exchange. Virtually every new coin or token is available here. You can even trade small-cap coins. Their prices are usually "pumped up," which means they skyrocket and then plummet. It is good for you if you are a trader. I regard this exchange as one of the safest.

How to Get Registered on a Cryptocurrency Exchange

There are many cryptocurrency exchanges available that everyone has a chance to find the one that best suits their needs. With that in mind, I want to clarify some issues regarding registration on cryptocurrency exchanges.

When getting registered on an exchange, I definitely recommend turning on two-factor authentication (identification). Even if you are already registered, I recommend that you use this.

I advise you to create an account on at least two exchanges; if something happens to one, you will be able to close the orders on another one. Be sure to use Bittrex if you decide to trade altcoins. Select another one by yourself.

Your password must contain at least ten characters: uppercase letters, lowercase letters, and numbers. Do not count on being able to remember it even if your password is the simplest combination of letters in the world. Therefore, be sure to write it down on paper and hide it somewhere safe. You can even make yourself a password tattoo, but you must write it down somewhere, either on your body or on paper! The second rule is not to use this password anywhere else except for the exchange. You must have a separate email for each account on the exchange. If one of your email accounts is hacked and wrongdoers try to get into your account on the exchange, your other account will remain safe.

Do not forget to check your email when depositing or withdrawing money from the exchange. These transactions must be confirmed via e-mail.

Last but not least: if you take the risk of trading on the cryptocurrency exchange, you need to gear up for the fact that a hacker (or even an exchange owner) may steal your money at any time. Any exchange may also crash. Therefore, it is very dangerous to hold money on the exchange. I advise you to only keep money you trade with on the exchange and to store the rest in cold wallets.

CHAPTER 9. DIGITAL CURRENCY EXCHANGES

If you exchange digital currency and it is important for you to do it with the lowest fees, you will have to resort to services of digital currency exchanges (DCEs).

There is also the possibility of a live exchange in person, but this option is not safe. There have been a lot of cases when a person has met up with another person to exchange Bitcoin but then gets beaten up and robbed. This risk can be minimized by using only the most popular DCEs, (the ones with the best reputations). You can find more information on the internet about the most reliable DCEs.

If you still prefer to take the plunge and try your luck in a physical, digital currency exchange, you should know that even stolen money can be tracked. A person who changes money for you usually has his or her wallet installed to a mobile phone. When he or she transfers a specific amount of Bitcoin to your wallet, you can immediately go to the Blockchain.info site and check whether the transaction has really been executed.

You already know that all transactions in the Blockchain network are open so that you can track them. To do this, enter the number of your transaction on the phone and wait for it to be processed. Once it is processed, you can be 99% sure that the money is in your wallet.

Where Can You Find Safe and Reliable Digital Currency Exchanges?

First of all, I recommend that you familiarize yourself with the LocalBitcoins site. Here you will find people who are willing to meet with you and make an exchange for cash or by bank transfer.

A reliable DCE should have a website a ratings and reviews section, support chat, and the amounts of currency available for conversion. For example, you can see that this DCE offers a particular amount of Bitcoin, Ether, and cash dollars.

Another good characteristic of digital currency exchanges is the possibility to reserve the price of a currency for a certain period.

As for the exchange fees, they range quite dramatically from 3% to 30%. Average fees on a calm market are 7%. When the market tumbles or perks up, the DCEs also raise fees.

CHAPTER 10. HOW TO EARN CRYPTOCURRENCY

Before you're ready to burst into the cryptocurrency world, there is one more thing you need to know: how to earn cryptocurrency.

There are several ways to earn income in the cryptocurrency market. First, you should carefully familiarize yourself with them and choose the one that suits you, depending on your lifestyle, financial standing, preferences, *etc.* You can even choose several coins and combine them skillfully. In such a case, you will have a better opportunity to diversify risks and earn more.

Nowadays, there are the following ways to gain profit in the cryptocurrency market:

- Long-term investment
- Trading
- Mining
- Participating in ICO
- Re-selling cryptocurrency.

Here is a brief overview of each of these methods. Then I will give you a more in-depth look at each strategy.

Long-Term Investment

If you choose this way of making a profit, then you should have:

- A period of 1 to 2 years (invest a sum of money you can easily "freeze" for several years)
- A substantial amount of money (if you can invest in parts only, this too can be viewed as a long-term investment)
- Readiness for risks (which are higher than in the banking sector).

You make money through significant capital gains in the long term.

Trading

If you prefer this way of making a profit, you should have:

- Time (not the time for waiting as with long-term investment, but free time you can devote daily to trading)
- A small amount of money
- Perseverance and diligence
- Mechanical intelligence (as you have to work with numbers and charts)
- Possibility to stay tuned and monitor developments.

You make money by beating the market and taking risks (the higher the potential gain, the higher the risk),in order to potentially make money quickly and gain recurrent income.

Mining

Mining is the process of producing new cryptocurrency or bitcoins. If you are interested in this type of making profits with cryptocurrency, you should decide whether you'd like to be a hobbyist or dive into professional mining.

For *hobbyist mining*, you need to have:

- Mechanical intelligence (or a friend/advisor with mechanical intelligence)
- Startup capital (usually up to six mining farms)
- Readiness for technical problems and interruptions
- Protection of equipment from external factors (pets, children, etc).

You make money through a small but constant profit in mining. You also enjoy the opportunity to not only to trade in the market but to manage it.

In case you decide to do *professional mining*, then you need the following essentials:

- Mechanical intelligence (or a friend/advisor with mechanical intelligence)
- Startup capital or investments
- Premises (with electricity, ventilation, cooling installations, security)

- A team
- Readiness for bearing responsibility (for equipment, premises, team).

By running your own mining farm, you can make a stable and constant profit; You will also manage the market, not just trade in it. In addition, this gives you the chance to own something real and sell your business in the future.

ICOs (Initial Coin Offering)

To date, a lot of people participate in so-called ICOs (Initial Coin Offering). Roughly speaking, this is another interpretation of a crowdfunding model, similar to an IPO. The participants finance the development of a project in return for future benefits but without any guarantees. ICO is the issuance of coupons or tokens by a certain project to be used to pay for site services with cryptocurrency in the future.

ICO has a lot in common with a venture fund, *i.e.* an investment fund focused on working with innovative enterprises and projects (startups). A venture fund invests in securities or shares of enterprises with high or relatively high risk and expects extremely high profits. As a rule, 70%-80% of such projects do not bring returns. However, the profit from the remaining 20%-30% makes up for all losses.

If you think you are brave enough to sponsor a project without guarantees, first ask yourself:

- Am I ready for extra risks?
- Do I have a capital cushion for diversification?

You make money with ICO by taking big risks with the chance to make super profits.

RE-SELLING CRYPTOCURRENCY

Re-selling cryptocurrency has a lot of resemblance to trading. Here you also make money on currency fluctuations, but unlike trading, it's about the game between the wholesale and the retail price rather than the market or exchange price.

If you are interested in this way of making a profit, then you:

- do not take risks
- operate in small sums only
- understand that income depends on your turnover.

So far, we have a superficial knowledge of the basic ways of earning income in the cryptocurrency market. Now let's dive deeper into each method.

Making Long-Term Investments: A More In-Depth Look

You may recall the first principle of Warren Buffett: never lose money. It is especially true for long-term investment as the most important thing here is to keep the money, not to lose it.

Let's outline the main principles of making a long-term investment:

- Do not lose money; reduce risks
- Do not make a fuss
- Reap benefit in future

This is how your cryptocurrency **portfolio** should look like for making long-term investment.

The blue color is used to indicate more moderate and stable positions, which can be less profitable. The green color shows more risky yet more promising positions. Blue coins are the currently more stable currency.

Why do we mark this currency blue? We do so because there is a strong possibility this currency will grow in value. Bitcoin takes a leading position here as it is still considered to be the most reliable currency. Besides, its price will only grow. Why am I so confident? Here is my answer:

- The entire cryptocurrency turnover is now carried out through Bitcoin
- This currency is accepted everywhere
- Swift development of the project and the system as a whole.

The dollar is also marked in blue because you should always have "quick" money in your portfolio, with which you can operate quickly and exchange for cryptocurrency.

Now let's look at the green coins in the portfolio.

If the previous currency can be viewed as almost unwaveringly stable, then the situation with green coins is constantly changing. The currency on the green list looks risky at the time of writing this book. However, keep in mind that everything can change the day you read this.

Judging from my observations, the formation of a portfolio depends on a person's age. Is it strange? Not at all! The younger you are, the riskier your investment portfolio can be. We do not mind being on thin ice when we are young, but the older we get, the less we can take risks.

What else influences your investment portfolio besides age? The size of your capital. If you have a lot of capital to invest, it's easy to calmly risk a small sum of money. But the more money you have, the more stable your strategy should be.

The last factor that influences the formation of your portfolio is the size of your financial cushion. If it is small, then it is better to reduce risks.

At the same time, trite as it may sound, I want to recall one more basic investing adage: buy low, sell high! When the coin price tumbles, it is often the best time to buy it. But at this very moment, we start to feel greedy, thinking the price of a coin may drop even lower.

This problem is solved by applying a "**moving average**" strategy. It can be simple or exponential.

Let's look at this example of simple moving average.

For example, you are very interested in a coin that costs $100. One fine morning, while drinking your coffee, you decide to buy 14 such coins. So you spend $1,400. In a couple of weeks, the coin price reaches $105. You sell these 14 coins for $105, getting $1,470. You have earned $70 in profit.

But what if you decide to buy these 14 coins, not at once, but one coin a week?

However, in one case, an average coin price is $100, and in another case, the price is $91. Accordingly, in the second case, you spend $1,274 on 14 coins and sell each of them at the same price as in the previous chart. Your profit is $196. How did that happen? We just chose a different approach to the situation.

In applying a simple moving average strategy, you do not need to constantly look for the perfect moment to enter the market or be afraid that the rate will plunge even lower. You just buy a certain amount of currency every month. You keep buying whether the market falls or grows. In fact, this strategy is very hackneyed but truly efficient for long-term investment.

In general, the simple moving average strategy allows:

- reducing the average price of purchase;
- buying more shares for the same money;

- reducing losses in the falls and emerging from losses more quickly on the rise.

The exponential moving average strategy allows the increase of your profits. Here everything works this way: when the price stays straight or goes up, make the usual purchase; when the rate goes down, you buy some more.

It is worth noting you can also use this method for short-term investment. Here you expect a coin's price to grow in the near future and then sell them. When using this strategy, I recommend paying attention to altcoins, especially those in the formation stage.

It is very important to **take profit** in the short-term strategy. For this, I recommend using the **50/50 strategy**. Suppose you buy a coin for $1.50. Later, its price reaches $3, and you sell 50% of a coin and break even. Everything is simple.

Let me give you a few tips.

If you are a fan of stability over risk, then give preference to coins that initially cost more. And if you dream of potentially explosive profits, then pay attention to cheap coins. **The biggest profit is made on cents.** After all, no one will argue that $1 will grow tenfold much faster than $100.

Finally, remember that the cryptocurrency market is constantly changing. Therefore, you need to monitor your investments hourly to stay up to date and make wise decisions.

Trading: A More In-Depth Look

Having read the word "trading," you might assume that not everyone can master this profession. You are right to some extent. However, you will succeed if you know a thing or two about trade, take an interest in news and events in the economic sector, and keep track of the exchange rate. You may even learn to make some financial forecasts.

In fact, trading in the cryptocurrency exchange does not differ much from trading in the stock exchange. It's just important to know how to analyze charts, make well-considered decisions, not to pander to fear, rumors, or emotions, and most importantly, to be ready to lose everything.

Trading in the stock market is all about exchanging currency pairs. Each transaction party dictates its terms, and the second, respectively, decides whether to accept them or not.

On the cryptocurrency exchange, you work only with orders to buy or sell. By placing an order, you set the price lower or higher than the available price. You state how many coins you want to buy and quote your price. If the market reaches your quote, the order is filled.

Let me explain when you can place an order on the cryptocurrency exchange. Let's say you read in the news that Bitcoin prices will fall and then grow again. It's the time for you to invest. I want to mention, though, that there is no exact formula which explains what coins you should invest in and when. The most important thing is not to invest at the stage of prolonged fall of a coin. On the contrary, it is good to invest in coins at the beginning of growth after a drawdown and, of course, in stable-growing coins.

As I have already mentioned, look through coinmarketcap.com to choose coins for trading. Also, keep an eye on the news about the release of new coins and track the price charts of main coins. Another important point is that you need to know the right time to buy or sell in order to make a profit. If you make a mistake, you have every chance to turn from a short-term investor into a strategic long-term one.

Do not forget about the exchange fees. I also want to remind you again: invest only the amount of money you are prepared to lose. You must in no case borrow money to trade with in the market. You will suffer inevitable losses in this market.

A critically important aspect in trading is security. The issue of security is crucial in any field that deals with money. All the more in the cryptocurrency market, where the amount of your investment portfolio can reach several tens of thousands of dollars, it would be irresponsible not to think about security in advance. You must research all possible ways to protect your finances.

What is the most convenient way to steal your cryptocurrency? The answer is: to steal it from the stock exchange. Let me share an example. In the summer of 2017, the largest cryptocurrency exchange in South Korea, Bithumb, reported billions of losses as a result of hacking. That's why I believe that keeping money on the exchange is risky.

The advantages of hot wallets, which are kept on the exchange, are accessibility and fast operation. You can easily access them as they are connected to the Internet. However, hot wallets have the huge drawback of vulnerability due to their connection to the Internet.

I won't elaborate any more on the topic of trading here since I devote a separate chapter to trading as one of the most widespread ways of making a profit in the cryptocurrency market.

MINING: A MORE IN-DEPTH LOOK

The word *mining* came into use precisely on the analogy of gold mining. Mining is the process of producing new cryptocurrency or Bitcoins. Owners or operators of mining devices are called miners. Often, the term *miner* is used to denote the computing device needed to spot Bitcoin (or another cryptocurrency) in the network.

Mining occurs by figuring out the digital signature of a block. A block in Bitcoin network is an array of data containing the information about transactions that hit the network after the previous block was created. A network participant, who has figured out the digital signature, is rewarded with cryptocurrency. At the same time, to get a "gold bar" in the form of a precious generating transaction, a miner needs to sift through tons of "dead rock," *i.e.* hashes unfit for the block.

Thus, each new block contains a digital signature which is formed on the basis of the previous block. The blocks adhere to each other and form a chain of blocks called **Blockchain**.

Most people think mining is money derived from an electrical outlet . But in fact, it's a great piece of painstaking work, which is rewarded with money by the system itself. Despite the fact that the client code and the cryptocurrency protocol code are fully open, creating new coins is a complex and expensive process. For example, you cannot generate more bitcoins than planned by the creator of the technology. To get new coins, you need substantial investments in equipment, premises, cooling systems, electricity, *etc.* That is why Bitcoin is called "digital gold" and is depicted in the form of gold coins.

Let's now answer the following question: Why does cryptocurrency need miners? There is a misconception that miners initially generated cryptocurrency. In fact, miners run the essential functions of the network:

☐ Confirmation of transactions

☐ Protection of the network from entering false information (fake transactions and blocks)

☐ Protection of the network from various types of attacks

☐ Maintenance of the network decentralization.

That is why the more mining devices in the network, the better the Bitcoin network is protected from any attacks. At the same time, the shutdown of part of its computation capacities will not halt transactions in the network. The network will continue operating if at least one miner works.

There are several types of mining. A home farm is one of them. It might look like this.

To date, specialized *ASIC* or *GPU* (graphics processing unit) chips are used for mining. ASIC is designed specifically for mining. It is a "soulless robot" which is tooled for extracting crypto coins. As for GPU mining, an ordinary farm usually consists of several video cards of one model, a forceful power supply, a motherboard with several PCI-Express slots, and processor with a cooling system. There are certain algorithms upon which coins work. You should select video cards depending on these algorithms.

What should you choose for mining: ASIC or GPU?

ASIC is easy to service and convenient to install. However, it is designed only for certain coins, and you will not be able to switch between them. If you choose ASIC, you will have to mine successfully enough to pay your electricity bills because you'll need a lot of electronic equipment.

As for GPU, unlike ASIC, you will be able to sell your video card easily and at a profit if you decide to give up mining at some point.

In my personal opinion, you need to have both ASIC and GPU for mining. If one of the devices fails, you can use the second one.

There is another option for people who do not want to get down to assembling a "money machine." You can mine remotely using shared processing power. This type of mining is called **cloud mining**. However, you should beware of the dangers of cloud mining. First, the payback period is quite deceptive. Second, you cannot control expenses and profit. Moreover, all the equipment is not in your personal property.

If you still decide to entrust mining to other people, then pay attention to the following things: you should not be given an enormous profit forecast, and you should be provided with the possibility to get in touch with technical support. I also advise you to read reviews and monitor the mining forums in regards to these companies as well as investing small amounts of money for testing purposes before investing more money.

In mining, like in trading, you should obey certain rules of security. The first rule of the so-called "mining club" is not to tell anyone where your farms are located. Is it a trivial tip? You may think so. But remember that you must be the only person to have access to the premises with your mining equipment.

Other security principles are as follows:

- Install the system to monitor indicators (temperature, etc.) as well as video surveillance and a fire fighting system on the premises
- Check the electricity, power, and earthing
- Monitor device heating and humidity on the premises
- Don't skimp on consumables as this may lead to a sad outcome.

By and large, mining does not suit everyone because you need to be able to negotiate with equipment suppliers, collect and configure the equipment yourself, and further maintain and ensure its security.

In fact, the topic of mining is quite profound, and I can talk much more about it. For example, I could tell you about the characteristics of power supplies, motherboards, and cooling systems; how to choose your operating system; what kind of cards are worth buying; how to work with suppliers and even share the secrets of payback. You can find the answers to all these questions in a free bonus book entitled *Cryptocurrency Mining.*

ICO

An ICO (Initial Coin Offering) is the initial placement of tokens on the blockchain. It is a specific crowdfunding model. ICO participants finance the development of a project in return for future benefits. ICO is often compared to IPO. But unlike IPO, participants in ICO often do not receive a stake in the company and cannot influence the internal management decisions.

It is believed that the first ICO was held by Mastercoin in June 2013. The Bitcointalk forum announced the launch of crowdfunding for the project. As a result, ICO Mastercoin collected more than 5,000 bitcoins.

However, ICO Ethereum, held in 2014, was perhaps one of the most powerful ICOs. Tokens then were sold for $0.30. In two years, their value on the exchange increased up to $12- $13, and in June 2017 they already cost $390.

There are a lot of ICOs nowadays. The new projects, which assure their prospects and look for investors, crop up almost every day. The more such projects are created, the less of them will be successful. A lot of ICOs are originally set up just to fleece people. The number of such projects will only grow as people become increasingly interested in making easy money through ICOs.

These days, the developers of these startups have even found a way to get the support of people who do not have the money to investing ICOs. The **bounty program** was invented just for them. It is an opportunity to get your share of the project tokens without investing your own funds. For this, you just need to popularize the project.

To get a bounty, you may:

- Promote the project (via social networks, email databases, campaigns on Bitcointalk forum) so that potential investors learn about it
- Translate information about ICO into other languages
- Look for bugs in the project
- Design logos, booklets, and other information (sometimes on a competitive basis)
- Design mobile applications, wallets, or other software add-ons.

If you have the skills for doing at least something from the list above, you have every chance to earn a bounty!

As you've probably already grasped, there are many scammers in ICOs. Let's talk about scam ICO. Just google "scam + history," and you will see a large list of various services that in one way or another have decided to play on the trust, or carelessness of others. The fields are the most diverse: from energy to the banking sector. Though ICO has existed only since 2013, this field has seen a lot of scams as well.

How can you tell if an ICO is a scam? Here's what to look out for:

- The ICO project creates only digital currency, which is positioned as a universal means for international payments without indicating its specific advantages
- The ICO project is described as extremely large-scale, though it does not present a specific model of profit distribution
- The ICO project developers are throwing every effort to promote it and say almost nothing about the code, which might even be closed (not open-source).

Scammers usually develop a platform without demonstrating any results of work. Sometimes the main developer is notorious in professional circles or has already been involved in the crook business.

It is worth mentioning that there are several examples of successful ICOs. The success of an ICO is dependent on a number of conditional things. The conventional indicator of success is the amount of raised funds. In this regard, the leader (at the time of writing this book) is the Bancor project, which raised 396,720 ETH in less than three hours in June 2017. An example of another rapid crowdsale was the innovative Brave browser campaign, which raised $36 million in the very first few minutes of the project. Furthermore, the projects Storj ($30 million in less than a week) and Aragon (which collected the required 275,000 ETH, about $25 million, in just 15 minutes from the start of the ICO) joined the ranks of successful cryptocurrency crowdsales in May 2017.

There are also the following projects: MobileGo ($53 million), Gnosis ($12.5 million in 10 minutes), Blockchain Capital ($10 million in 2 hours), Aeternity ($2 million in the first few hours), and, finally, the Status project (about $100 million) which crashed the Ethereum network. And we can't forget about the legendary The DAO project, which collapsed and gave birth to Ethereum Classic in summer 2016.

The success of an ICO largely depends not only on the idea but on a smart PR campaign as well.

You may ask: How are the investors' funds protected in such projects? My answer is very simple: The funds are not protected. By and large, the only safety guarantees are the reputation of the project, *i.e.* the reputation of the people unknown to you.

Why do people still invest in these projects considering all these factors? Here are some reasons people take the risk of participating in ICOs:

- Desire to make money on a trend
- Desire to assume the role of an investor
- Desire to keep up with other investors
- Belief in a startup.

That's why it's very important to carefully choose an ICO to invest in as there are no laws governing the conduct of ICOs and protecting you from scammers. It is not uncommon in ICOs, like in crowdfunding, for a project to never even get to the stage of product implementation.

If you are still interested in a certain ICO and ready to invest in it, pay attention to the following key points:

- Project's whitepaper and road map
- Essence of the product (feasibility, demand, is there a real problem which the project solves?)
- Connection of project with Blockchain (whether it is needed or far-fetched)
- Project team (experience, their history)
- Main persons behind the project (experience, their history)
- Substantiation of the project's purpose (what money is needed for).

You should also find answers to the following questions:

- what will drive the demand for these tokens/cryptocurrency?
- what will provoke a large turnover of coins and what can be acquired for them?
- whether you will be able to exchange them for goods or sell them?

At the same time, when you analyze an ICO as an investor, you should understand whether the project can independently solve the problem or task it's designed for.

Moreover, keep in mind the following tips:

- Choose a project in which you understand at least something
- Invest rationally, don't agonize over it, and do not put all your money on one project
- Always be careful. Even if you are lucky with one ICO, you are not guaranteed against loss forever.

You can use icorating.com to gain insights on the companies running ICOs. Icorating is an independent agency where you can find the lists of different ICOs, both closed and upcoming. There is even a list of ICOs considered to be blatant scams. The icorating.com experts analyze the following ICO features:

☐ Business model (its relevance, merits ,and flaws)

☐ Market niche (prospects and dynamics of development of the chosen niche for doing business)

☐ Team (business experience in the traditional market segment, Blockchain industry, experience in Blockchain development)

☐ Competition (level of competitive pressure in relation to companies with similar business models from the traditional market segment and Blockchain economy)

☐ Technical background (availability and quality of a prototype or source code)

☐ Analysis of feedback from the community

Having reviewed the preliminary assessment of ICOs by this agency, it will be easier for you to analyze the project you're considering in order to make a final decision on whether to invest in it or not.

Another resource for ICOs is icotracker.net. Here you will find a list of upcoming ICOs. I also recommend icobazaar.com, icocountdown.com and cyber.fund/radar.

The ICO market now resembles the boom of dotcoms (1995-2001). People today remember the time when companies like Apple, Amazon, or Google arranged IPOs and wonder at the kinds of fortunes they could have made if they had invested in those companies in back then.

The ICO market gives you another chance to become a part of these new cyber success stories. Some ICOs have the potential to grow dramatically over time and to recoup the initial investments manifold. Still, according to some experts, most projects in 2017 may turn out to be frauds. It makes sense, as it now easier to invest thanks to the virtual economy. People invest their money practically sight unseen in any company with a more or less decent presentation. Such a thoughtless approach plays into the hands of scammers.

CHAPTER 11. ICOS

Here we'll go through the basics of ICOs (Initial Coin Offerings), compare them to IPOs, how to identify a good ICO versus a scam, and how to participate in one.

ICOs vs. IPOs

All stock investors are familiar with the abbreviation IPO (Initial Public Offering), which means the initial public offering of stocks. The buyers of stocks become co-owners of the company, and the company itself gets money from sales, usually exceeding its annual returns. Recently, a similar abbreviation has come into existence in the cryptocurrency community: ICO, or Initial Coin Offering.

An IPO is a legitimate (legally speaking) operation. A company with an existing and maturing business structure launches an IPO to attract funding. Raising funds through selling shares is the cheapest way to do this. At the same time, launching an IPO is a very expensive process. For a company to enter the exchange, it must meet many requirements, including the volume of capitalization and undergo an audit. In addition, the company should attract underwriters who will sell the company's shares. Then there's getting the word out, attracting money and investors, and then the company will start to trade its shares on the stock exchange. That's why it is a long and expensive process.

Also, a deficit is deliberately created within the framework of an IPO, and only a limited number of the company's shares enter the market.

The creators of ICO campaigns also declare that the number of coins is limited, and because of this, many people press the panic button, thinking if they do not buy these coins now, they will have to buy this currency later but at a higher price.

Both IPOs and the ICOs are risky ways to make money, but they can also be very profitable. People participate in ICOs hoping to make a quick buck. And as soon as these people earn their money back, they immediately withdraw them. That's why the coins first zoom up sharply and then go down in price when people start to sell them. This is one of the strategies the speculators use.

In contrast to ICOs, such speculations are rather rare in IPO because of the so-called "lock-up period." People, who bought shares before the IPO cannot sell them until three months later. With ICOs, you can buy coins even during the pre-ICO stage.

Let's summarize why so many people take the risk of investing in ICOs. Some invest money in these projects simply to support them, while other people will quickly buy and quickly sell, and the third group of people makes the investment because they read about the project somewhere or, even worse, want to duplicate the success of their friends.

How to Distinguish Between Promising ICOs and Pyramid Schemes

All companies that launch ICOs are startups that do not even have a so-called Minimum Viable Product (MVP). They have only an idea. Most of these startups are increasingly showing signs of pyramid schemes, namely you are assured that if you invest money at the first stage your investments will be doubled already at the second stage and will grow fourfold at the third stage. In other words, investors are offered a sort of game, so that they get the impression of being able to get rich quick.

Let's study how an ICO is conducted.

First, the companies run a so-called pre-ICO, the token sale event before the official crowdsale or ICO campaign goes live. This event is typically held to raise funds for marketing, particularly to pay for a PR campaign during the ICO. In other words, it sounds like this: "We raise money for advertising to arrange an even greater promotion later and collect money again."

In addition some ICOs have will stop to declare that a number (more) of funds is needed for the actual project implementation, so the largest possible sum is collected from the market. This is how sometimes people create a company with a capitalization of billions of dollars.

IOTA currency is a bright example. It is a very scalable currency that can support many transactions. To date, IOTA is valued at almost $2 billion. But there is one catch. The company promises nothing and owns nothing (except for the site of course). That is, by and large, people value the design of the company's site at almost $2 billion! It's not a good investment, is it?

Using the example of the IOTA currency chart, you can see what all the latest ICOs look like to date.

I admit that some ICOs zoom up in real value greatly, and those who have managed to participate in them can really make a good profit. However, all such ICOs still have the one and same fate, and it can be compared with an elevator: when it is full, it cannot go up. So there is a huge amount of speculative money in such ICOs; every price surge makes these people sell everything, close their positions, and thereby exert pressure on the stocks.

By saying all of this, I do not intend to make you believe that all ICO projects are deceptive. In fact, I just want to show you that most of these startups are very overrated and only 2% -3% of them survive.

To profit from ICOs, you must, first and foremost, understand why you're investing in it. So the reason for your transaction entry should be equal to the reason for the exit. If you participate in an ICO simply because you believe in a future product, then it is probably not the smartest decision. First, let the company enter the market and start to trade. If you still believe in its product, then you can always buy it, even at a discount, some time later.

How to Identify Good ICOs

The characteristics of a good ICO are as follows:

1. Chief executives and developers are public figures who have been active on their social media accounts for quite a long period
2. A clear understanding of project tasks and objectives
3. Competitors in the market are making a profit
4. Support from community opinion leaders
5. Deliver speeches at cryptocurrency events
6. Good evaluations by experts and rating agencies
7. Distribution of money after ICO period
8. Open source code on GitHub
9. Related information is available on Bitcointalk, Reddit, Slack, Twitter, Telegram, and YouTube.

When you have already chosen an ICO and decided to invest in it, now you should constantly monitor the information this startup releases. The best way to do this is to join their chat and follow the project on Twitter. Surf through their Twitter archive. Do not forget that you gave them your money, so do not be afraid of work and time spent on it.

You can keep track of ICOs on sites like smithandcrown.com, icoalert.com, and icotracker.net.

Now I will give you some tips on being wise when it comes to investing in ICOs.

I recommend that you buy pre-bankruptcy ICOs and just keep them in case one of them booms. To be frank, I have a separate list of all dropping ICOs, and I keep track each of them every day. I am ready to buy them for some disposal value. For example, 10% of the placing price. In this case, I will gladly make a purchase, for example, at $1,000, and thus my risk will be the same $1,000.

The next tip: many people think that if the price starts to go down, it is a sign of bankruptcy and you need to run away. And vice versa: if the price grows, you need to buy. In fact, everything works the other way round. **Falling market is the best opportunity for identifying strong and weak coins and tokens.** Track whether the volume of coin increases when it falls. If it increases, that's good. If the coin price drops and the volume does not increase, it means there is no demand for these coins, and a seller constantly has to lower the price.

In addition, do not forget that, currently, 99% of all ICOs are a scam. However, you may participate these ICOs too. You read right. You may do it to make a profit on greedy and ill-informed people. In a certain way, this also proves that ICO is a legal pyramid scheme, and your task is to enter it and leave it as soon as possible.

How to Participate in ICO

Suppose you found a good ICO. In order to participate in it, you need to visit its site. Go to the Tokens tab and click on Get EOS.

Afterwards, be sure to tick the box that says you agree to the terms of the project. Read between the lines: you agree to possibly lose your capital. After clicking Continue, you will be taken to the page which explains how the project will raise funds.

I recommend that you carefully review all the figures, especially the ones about capitalization. If the ICO capitalization is about $500 million, it's okay. However, if you multiply the value of a token by the number of all tokens known in advance and get the capitalization of $5 billion, this gives pause for thought. You can also compare capitalization of the ICO you are interested in with a previous ICO. When comparing, pay attention not to the value of one token but to the capitalization of the whole company.

Study the Schedule section carefully to find out how long the ICO will last.

The ICO tokens you purchased appear very quickly on the hitbtc.com exchange. Here you can also find tokens offered by ICOs.

In any case, if you fail to figure out how to participate in the ICO, the sites of any ICO always publish detailed instructions (sometimes even a video). There you will also find the information on how to transfer money to participate in the ICO. But remember that this money can be transferred only from your wallet, not from the exchange.

So that's how ICOs work and how I recommend you invest in them. No matter what, remember this: Don't stay in one ICO for too long. Leave it as soon as you can make a small profit.

CHAPTER 12. TRADING

Here we will cover the basics of placing orders on the exchange as well different kinds of trading and how to analyze chart patterns.

Placing Orders on the Exchange

To open and close positions on the exchange, exchange orders are used. Depending on the type of execution, there are market orders and pending orders. Pending orders, in turn, can be limit orders, stop orders, or stop limit orders. Each type of order can be used to buy or sell some kind of asset.

I recommend using only limit orders, but we will discuss this point in detail a bit later.

Let's consider an example of an order on the Kraken exchange. An order may look different on different stock exchanges, but its essence remains the same.

The easiest way to buy or sell something on the stock exchange is to place a market order. This order is executed at the current market price immediately after placing it on the exchange if there is a reverse limit order for it. For example, to execute a market buy order, a sell limit order is required. To execute a market sell order, a buy limit order is required.

To execute a market order, the best one is chosen from all the available limit orders. Therefore, the market buy order is executed at the ask price, while market sell order is filled at the bid price.

When dealing with this kind of order, you do not need to specify the price, just the volume, *i.e.* how much you want to buy. However, there is a big problem you can face - slippage. What is it? It is the difference between the expected price of a trade and the price at which the trade is actually executed. Slippage may occur during periods of higher volatility when market orders are used and also when there are large orders there may not be enough interest at the desired price level to maintain the expected price of trade. It is for this reason that I do not recommend using market orders at all.

Before proceeding to other types of orders, let's look at the terminology mentioned above: bid, ask, andsSpread. A bid is the price a buyer of the cryptocurrency is willing to pay. Ask is the price a seller is willing to accept. Spread is the difference between these two prices.

In market relations, the buyers set the price they want to get an asset for while the sellers set the price they want to sell an asset for. Put simply, it is bargaining: a buyer sets a lower price, and a seller insists on a higher price. It is the situation observed on the cryptocurrency exchange.

So the orders of buyers with the specified prices stand in a kind of line. To be the first in this line, you must offer the best price. At the same time, sellers with the ask price also line up. The first in this line is also the one who offers the best price. Each price has an indication of the volume a buyer wants to buy or a seller wants to sell. Traders call such lines of pending orders on both the bid and ask side *market depth*.

Here's an example. Let's say you place a market order to buy coins for $5,000, but the value of available coins is just $1,000.

The price you pay for the coins at the time your order is executed will be in between these prices. I should also mention that the exchange is also artful and is always happy to earn something. If you place a market order, then be 99% sure you will never get the very first ask price. As a result, you've sent your order, and then, for example, get a price of 244.78.

By the way, in this situation, you can use the easiest way to make a profit on the exchange without taking a risk - make the spread, *i.e.* the difference between the bid and ask prices, wider.

I recommend using a limit order to avoid similar situations. A limit order is an order to buy or sell a certain volume of an asset at a specific price. Here we consider two parameters: price and volume. For example, you place an order to buy coins for $5,000 at a price of $243.47. The first option: you can place a buy limit order, for example, in the bid line. So you join the line of buyers waiting until their orders are filled. Another option: you can come to the sellers and make a deal at their price, but one which is not lower than yours.

However, it may happen there is no sufficient volume to fill your order. What should you do to get the needed volume? You should evaluate your order with regard to the overall liquidity of market depth. That is, you need to examine what orders have been placed. Theoretically, you have a chance to gather the first best positions regarding price and volume, but there is always a possibility that someone had already placed an order before you and took this volume because the information is sometimes displayed with a delay. If you still want to fill your order (not using a market order to avoid large slippage), you simply place a limit order at a price where your order will be definitely executed, not at a price you wish it were executed. This means that you are very likely to fill your order at a price offered by a seller. If these orders are in the market and you've managed to hit them, then you increase your chances of filling your order having offered a little bit outsized price.

Now let's touch on the topic of stop orders. A stop order is an order to buy or sell a stock when its price surpasses a particular point.

Let's say you set your exit price at $200. Thus, once the price reaches $200, your stop order turns into a regular market order. However, I do not advise you to use this type of order.

There is also a stop-limit order that combines the features of a stop order with those of a limit order. A stop-limit order requires the setting of two price points. The first point initiates the start of the specified action, referred to as the stop, while the second represents the outside of the investor's target price, referred to as the limit.

In my opinion, a stop-limit order is even worse than a stop order. It does not guarantee the order execution, even if a price reaches a given stop price (because the limiter might not work). For this reason, I do not recommend using stop-limit orders to close position with a view to limit losses (stop-loss). In addition, not all brokers accept this type of trade order.

In other words, it seems that stop orders should protect you from losses because if something suddenly starts to fall, you can close the position. In fact, it doesn't work that way, and stop orders will be of no use. You might think that you place a stop order, it will definitely turn into a market order when the price reaches a specific level and your position will close. However, when you use a stop-limit order, it is logical that your stop turns into a limit. If there is a price in the market to execute your limit order, it will be executed. However, if the market does not offer such a price, you will simply stand in line and wait. Thus, I believe it does not make sense to use a stop-limit order.

There are such orders as GTC and Fill or Kill. A GTC (Good Till Canceled) order remains active until it is either rescinded by the investor or the trade is executed. In this regard, such an order is very convenient. Fill or Kill order is a limit order variation. If in the previous situations you placed a limit order, you may get only part of your desired volume, but with a Fill or Kill order, your order must be filled in its entirety or cancelled.

Margin Trading

The essence of margin trading is very simple: you trade borrowed money. When you buy on margin, you pay a portion of the stock price (called the margin) and borrow the rest from other players offering money. The balance of your margin account is used only to serve as this borrowed money and to cover loan costs if necessary. In other words, margin trading allows you to trade the amount of money you don't actually have. The borrowed capital you trade is called *leverage*.

Let's say you have $10,000. Relatively speaking, it is your margin. Say you decide to trade with 1:4 leverage. This means you can make a $40,000 transaction. Your margin is 25%. You buy for $40,000, and if the purchased currency starts to grow, your profit becomes four times larger. but if it all starts to fall, your losses are also four times larger. Accordingly, if you buy cryptocurrency for $10,000, you may lose all your money only if your asset (for example, Bitcoin) drops to zero. If you buy cryptocurrency for $40,000 with leverage on the margin account, you will lose all your personal money ($10,000) when the asset falls by 25%.

Or let's imagine another situation: you keep margin position in the Ether when a flash crash occurs. In this case, your position will be closed by the exchange itself since the Ether fell at that moment by more than 25% As soon as your loss is $10,000, the stock exchange will close your position of $40,000. That is, $10,000 is something like collateral for your trading.

Margin or leverage in any other market is a good tool for maximizing your profit. Almost all experienced traders use leverage. However, as we said earlier, the cryptocurrency market is highly volatile, so here the leverage can figuratively kill you. Why does it happen? Let's compare it with the stock market.

What can happen to make Apple shares drop, say, by 20%? I think nothing of a kind can happen. Therefore, it is largely safe to trade Apple shares with leverage as these shares cannot suddenly fall by even 10%. At the same time, you can always close the position in the stock market because the trade is strictly regulated here, and you can follow all the events.

On the contrary, the cryptocurrency market operates 24/7. Any developments are possible here, including the most illogical and inexplicable when the price can grow or fall by any percentage points. Therefore, if you trade with margin in the cryptocurrency market, you'd better use the leverage of 1:2. Trading with a higher leverage will resemble a Russian roulette. If you trade with leverage 1:4, your position is most likely to be simply destroyed.

Always be cautious: if you find an exchange with a suspicious interface or you are offered 1:20 leverage, this exchange must be hunting for marginal traders. It hunts for greedy, cunning, and, as practice shows, stupid traders who want to increase their profits twenty-fold immediately. Therefore, I will remind you once again: if you decide to trade professionally, you should undergo verification in the top exchanges. You should open accounts on at least two of them.

You should have two accounts for margin trading: the first one for your money and another one for margin. To start trading with leverage, you should transfer money from your first account to a margin account and trade from there.

You also need to know the following things about margin trading. Do you think you will be given money for trading for no more than a "thank you?" Of course not. You will be given money only at interest. If, for example, you use the leverage of 1:4 and want to trade not for the $10,000 you have but for $40,000, then you will pay a higher interest. It can reach 1% -2% per day. In the case we're considering, the interest accrues for $30,000. I remind you that you actually only have $10,000.

And, finally, here are some tips for beginners.

When trading in the market, buy the assets whose value has fallen dramatically. This is my main strategy now: to buy underrated assets or the assets whose price has tumbled.

To close a position, put an order in only in case you have managed to get a profit of more than 50%. It is very risky to wait for the larger profits, so close your position when you have earned your 50%. I also do not recommend taking profit of less than 10%. In my opinion, 20% is the minimum you should aim at. However, feel free to place sell orders at 30%+ if you do intraday trading.

Do not forget that the exchange charges commission fees. The less you trade, the higher the fees are. And the more your turnover grows, the more the commission fees drop.

Day Trading and Long-Term Position Trading

Let's focus on the schedule of your work on the cryptocurrency exchange.

Before you start to trade, you need to decide on a very important point, namely how much time you are ready to dedicate to trading. Of course, the best option is to sit at the computer for 5-6 hours every day, constantly closing positions and monitoring the situation. However, most people now choose another option because of the lack of time. They trade once a week. On a particular day, such traders look through all the news for the week as well as any charts and then decide what positions they should open or close.

A lot of people go with this option so here are some tips:

- Choose one specific day of the week which you will consistently devote to trading
- On this day, look through your portfolio – what currency has grown in value and what has fallen?
- Monitor the news

Make decisions depending the news and composition of your portfolio. If some asset has not grown or has even fallen in value, then close the position. If you have noticed something promising or falling in price, buy it. You should carry through these operations every day of the week you have chosen for trading.

By the way, I recommend using Blockfolio to make tracking your positions more convenient.

To recap, I advise choosing between two types of trading: daily monitoring with limit sell orders or monitoring once a week.

At the same time, some cryptocurrency exchanges, for example, Bitfinex, offer OTC trading. The abbreviation OTC stands for Over The Counter, which means over-the-counter transaction. This is a decentralized market without a central physical location where market participants trade assets through the dealers' network.

For example, if you have no time to follow the market and be fully engaged in trading by placing orders, you can use the services of this market. Your cooperation will be carried out via mobile phone or e-mail. So you tell a dealer that you, for example, want to buy 250 bitcoins. Having looked through the market, the dealer answers how much you should pay for this amount of bitcoins. A dealer will charge fees for this service.

Analyzing Chart Patterns

It is widely believed that a natural gift or instinct for trading is one of the key success factors on any exchange. However, it is misguided thinking. To make a profit, you primarily need to learn to analyze the market situation. The skill of reading the charts will allow you to make the right trading decisions and make money.

There are two kinds of chart patterns in the cryptocurrency market: line charts and Japanese candlestick charts. Here is some practical information on each of them.

Line charts work fine for medium-term trading as you do not pay attention to the minimum and maximum indicators. Such indicators are taken into account in the Japanese candlestick charts.

Japanese candlestick charts are the most common way of displaying price charts. Traders love them for their simplicity and clarity. Unlike line charts, candlesticks are more useful for decision-making in trading. They allow you to see the overall picture of the market and forecast price movements. So it is really better for traders to analyze Japanese candlestick charts, although I think that newcomers need to work with line charts. Even more so if you are going to make long or medium-term investments.

Each chart has its own time frame, *i.e.* a period you can look through. In the Japanese candlestick charts, the time frame is formed over a period of one hour, one day, one week. Candlesticks are usually composed of the body (black or white), and an upper and a lower shadow (wick). The body (rectangle) illustrates the opening and closing trades. The wick illustrates the highest and lowest traded prices of a security during the time interval represented.

We can see the opening price and the closing price. These two positions are usually marked with two colors. If the opening price is below the closing price, it means growth and is marked with a different color on the chart. If everything is vice versa, it that stock is falling.

It is a little more difficult to determine the time frame on a line chart. Usually, I see how the situation changes point by point.

What time frame should you analyze? Personally, I do not analyse intraday charts. The one-hour chart is the smallest one I work with. I mean the 1 hour + daily chart. It's also a good idea to analyze the weekly and one-month charts as the cryptocurrency market is quite young and you will be able to see the previous minimum and maximum prices.

What should you pay attention to? In technical analysis, all people respond equally to certain things. So every person who looks at the chart will equally see and say: "It has dropped here but grown there. It has started here, but it's now over there." Charts push people to act in a particular way, and they say: "Oh, the maximum price has updated, so I will buy." And then I notice that people tend to buy when they see round figures (for example, when the Ether reaches $400, or Bitcoin breaks $5,000).

I also notice that all people react to recurrent events. If you have already reached a peak, you hope you will break this peak again and go up even higher. However, the market for real assets differs much from the cryptocurrency market. The real assets market has its own algorithms and big players. There are also big players in the cryptocurrency market, but this market is still very young as it began to exist only in 2013. Yes, in 2013. Perhaps, some individual IT specialists or traders started to get interested in the cryptocurrency market back in 2011 or 2012, but speculators entered it only in 2013. Therefore, this market is very young, and it has not faced any crisis yet. Despite this, the market is continually updated, *i.e.* there are definitely certain cycles. Therefore, recurrent events are double or triple peaks and, accordingly, double, triple floors. That is a constant update of the minimum and maximum prices.

Market indicators also help to catch the mood of the cryptocurrency market. Personally, I do not use them, but I highly recommend you use these tools.

In a volatile market, people somehow pay attention to the **moving average**. What is it? As a rule, when you look at a chart, you have a median in your head that divides the schedule in two, and you can immediately determine the average price. This median halves the chart: everything above the line is expensive while below means cheap.

Thus, for a daily chart, I sometimes use a 50-day moving average as well as a 200-day moving average. You can also apply a 20-day moving average. However, I do not really need them as my eyes are trained enough to determine the average.

The most important indicator to be used is the **volume**. It shows the number of transactions for a certain period. If you use a candlestick chart, this indicator shows a number of trades for one candlestick. The overall volume is shown, *i.e.* how much was bought and how much was sold. Therefore, you can always determine the liquidity, *i.e.* to understand how much you can buy in principle.

Thus, I believe that any movement of yours must be backed up by volume. If the volume starts to grow, this is a good sign because it shows that more and more participants coming to the market.

Therefore, any strong movement (hitting a new high or low) must be accompanied by volume. Moreover, it must be accompanied with a **volume spike**. It is from this point, as a rule, that all movements start and end. If you see that the price of a coin is growing strongly, the volume is increasing, and a volume spike begins, it is a good opportunity for making a profit.

Another situation: if you see that a movement begins to fall and the volume, on the contrary, starts to increase, you'd better wait for the volume spike.

Now let's consider where you can follow the charts.

In fact, the difference between the Poloniex, Kraken, and Bitfinex exchanges is extremely insignificant. For example, the currency price can be $245 on Kraken, $248 on Poloniex, and $244 on Bitfinex. It seems there is a difference, but the charts are actually completely correlated and very similar. Therefore, it does not make much difference where exactly to follow them.

Let's now discuss when you should enter the trade.

You should enter the trade only with an eye on volume. The first reason is when there is a **range** (when the price stays within a particular range and does not move) + spike. You should buy when the price moves out of range and sell when the price goes down. The second reason for entering a position is reaching a new high and low (new high/new low + spike). And the last reason is the narrowing and widening of volatility. This method is also called a triangle. It does not matter what kind of triangle you have since the most important thing is a surge and growth of volatility.

It looks like this on the chart.

A volume spike confirms that this pattern will work indeed. And how is volatility narrowing? It testifies to the fact that someone wants to close the position.

I'm sure that you do not need to look for a lot of strategies to enter because the ones I've described will be enough for you.

You should also grasp that the best opportunity for buying is the so-called **retest**. There are different support and resistance levels.

A retest is the safest entry (in terms of transaction risk) as you have a reason for entering, a breakout, a confirmation no low will be reached and, finally, an exit. If this is accompanied with volume, everything is just perfect. That is, there are horizontal resistance and support lines and volume.

Now I want to give you my last tips.

Your entry should be put into writing. That is, you should define the clear reasons for entering in your notebook. For example, I buy because: all major currencies grow; my currency updates the maximum; my currency hits a round figure; good news has appeared about my currency, *etc.* Every time you do something, you have to take a pen, a sheet of paper, and write everything down. If it is not written down, you can deceive yourself. After all, only written rules will be obeyed.

Furthermore, do not confuse the technical entry (based on technical analysis and charts) and the fundamental entry, which does not imply a clear exit. It is a bit stretched in time and you exit only relying on the conditions of money management (we will talk about this later). For example, you have decided never to take more than 10% of loss in a position. That is your limit. However, technical analysis implicates a specific entry point and a specific exit point. Therefore, most beginners use technical analysis as it is simple enough to determine the entry/exit points.

CHAPTER 13. INVESTMENT STRATEGIES

Now it's time to take a look at some of the most important information I have to offer you, namely cryptocurrency investment strategies. Grab a notebook and a pen, and let's dive in.

7 Investment Strategies + 2 Perfect Strategies

The Way Of The Samurai

I call the first investment strategy *The Way of the Samurai* or *Just Buy Bitcoin*. So, following this strategy, you only need to buy Bitcoin.

This strategy is good in all respects. It is simple to implement, and you will not have to delve into the subtleties of economic or technical concepts.

You might make the biggest profit using this strategy. Nevertheless, this strategy has its risks, which I'll explain in a bit.

Follow The Money

Here I suggest that you buy a certain amount of different coins among the top 10 best cryptocurrencies for today. In other words, you will buy several market leaders. The coins, whose price will skyrocket, will compensate for the losses provoked by the failure of the other coins.

This strategy also implies huge risks. Nothing can protect you from the fact that seemingly stable coins thus far can considerably fall down in value next year. You should remember that right now there are no fundamental grounds protecting Bitcoin from failing, which, for example, may be provoked by the inability to withstand the competition of other coins. This is a very real possibility if only because Bitcoin is one of the technically weakest cryptocurrencies.

Therefore, if you do not want to become a hostage of circumstance, do not place your bets on just one currency as I did five years ago. I want to share this story with you. Let me tell you about the most expensive night of my life, which cost me $60,000.

So how did I manage to spend such a big sum of money in one night? It's actually quite simple. One fine evening, I decided to relax with my mates in a small town where my friend lived. We went to a bar, famous for the most delicious fish and stunning cocktails in town. Back then, I followed the price of the currency I bought earlier (Bitcoin) several times a day. I had noticed it was going down every hour. For some reason, I wasn't very upset when my cell phone battery died, and I could not check the Bitcoin price on the exchange. I believed the matter could wait until later.

I ended up losing more than $60,000 on the exchange during the three hours we had fun at the bar.

If I had been less dismissive of the dying phone and had access to the cryptocurrency exchange, I could have had the opportunity to sell the currency before incurring such devastating losses. I will definitely remember that night in the bar for the rest of my life.

Trading

Many people think of trading as analysis, patterns, and technical modeling, but the trader's work is actually a subjective assessment of what is happening in the market.

The first thing good traders should do, besides drinking a glass of whiskey (grin), is to predict what cryptocurrencies will grow in the market. After that, traders need to predict the best point for entering the market. What's next? They should continue making forecasts of when coin prices will start to fall.

However, the most important part of trading is having a colossal amount of experience and even a little bit of luck. A trader must also repeat all the above actions constantly. This is the problem because everything does not always work out well, whereas the risks have to be assumed all the time.

That is why, in my opinion, trading suits people with an engineering mentality, who have a lot of time, and, most importantly, are strong in spirit.

Mining

You only need computer equipment to implement this strategy. What's next? You plug it into a socket and wait for money to enter your pocket.

What do you get money for? As I previously mention, miners get rewards for providing the network with computing capacities. The network rewards miners with coins for their equipment and work.

If you want to learn how to get started in mining, make sure you download a bonus book, Cryptocurrency Mining.

ICOs

Nowadays, the cryptocurrency market offers more opportunities for active investors. This kind of strategy suits you if you have a lot of free time (and money), enough experience, and you want to make big and swift profits. Although this strategy is very profitable, it is also very risky, especially for beginners.

The principle thing of ICO strategy, which I also call *startups*. is as follows. The cryptocurrency economy has come up with a very simple way for people who have innovative and interesting ideas to fund their projects. They no longer need to pound the pavement, begging and convincing big companies of the profitability of investing in their idea. Now these inventors and developers simply share a proposal on their website. If you're interested in their idea, you publish the address of your wallet to receive tokens if this idea proves to be a success. The risk is that you have to invest money right now for a potential reward in the future.

In the past, the rules of the game with ICOs were quite simple. I give you one dollar, and you give me two tokens, Now some ICOs go to absurd lengths. Generally speaking, after have you invested in ICO, it may turn out that you can exchange their tokens only on Friday at midnight at the full moon and only if you are a mermaid with a blue tail. Maybe my example is obviously a bit of an exaggeration, but it perfectly demonstrates the opacity of the rules of most startups (ICOs).

If it was not very difficult to find a grain of truth among the variety of ICOs half a year ago (since only a few of them appeared in a week, and they could be easily analyzed), now new ICOs appear almost every hour, and you have no time to examine them all.

What would I advise a newbie who is tempted to try his luck with ICOs?

First of all, you need to have a good handle on the business niche which you are going to invest in. If you are savvy in it, you can analyze it and understand if it has prospects for success. I also advise you to listen to what experienced investors and experts say and pay less attention to what is written on the ICO's website. I would even say that you should not invest in any ICO that is not mentioned in at least three professional analytical reports from recognized experts.

PENNY STOCKS

If you know a thing or two about the exchange trade, the notion of *penny stocks* should definitely be familiar to you. Roughly speaking, these are shares of the company that nobody needs. However, these shares have one feature: they are so cheap that nothing prevents them from shooting up tenfold in value. Such a magical leap can occur because of good news, a little market manipulation, or any other reason.

So how can you take advantage of this strategy? You buy the oddest and more obscure coins and then wait and see how the situation will develop. During the year, the prices of some of your coins may skyrocket. If this happens, you should immediately sell this currency. If you bought fifty coins, a leap in the price of even one super-cheap coin will compensate for your entire investment portfolio.

This kind of strategy doesn't allow you to dive deep into the principles of doing cryptocurrency business. All you need to do is buy "garbage" and sell it in time. However, this strategy is not my prerogative as it resembles a casino. You need no skills or knowledge, only luck. You cannot in any way influence what happens or manage risks as everything depends on good luck.

Trust In Me

This strategy is right for you if you do not have experience in the cryptocurrency business but have money to play with. However, you need to be ready to take risks and entrust your money to strangers. The world of cryptocurrency business is regulated by no laws, so nobody can guarantee you that a company or person who promised to earn money for you will not take your money and take off to some warm country.

There are a lot of companies in the market that offer profit in return for trust. Some companies, for example, offer mining contracts to those who want to mine but don't have the capital to invest in all the equipment and space. This is called cloud mining. Another type of company is cryptocurrency funds. These companies don't know how to mine but are good at guessing promising coins. The third type is ICO mutual funds. These guys spend their time examining the startups in the market, which they will later invest your money in. Finally, the fourth option is trading. You give control of your money to someone else, who will do the trading for you.

Now that we've gone over these seven investment strategies, please realize that **I CONSIDER ALL OF THESE STRATEGIES TO BE IMPERFECT.**

Before you throw this book across the room, let me share two **PERFECT INVESTMENT STRATEGIES**.

In my opinion, there are two perfect strategies.

The first one revolves around you sincerely believing in the future of cryptocurrency. If this is you, then you should take on a few strategies of interest to you, mix them up, and start making profit that way: by diversifying your chances of turning a profit as well as your risks.

If deep down you don't truly believe in the future of the cryptocurrency business, then you should pay attention to the **second perfect investment strategy**: earn money through those who do believe in the future of cryptocurrency.

I call this strategy **Selling Shovels**. Some people will need mining equipment and premises, others will need funds, some will need information, and you can make money by fulfilling their needs.

CHAPTER 14. RISK AND MONEY MANAGEMENT

Money management is probably one of the most important concepts in investing. If you lack a competent approach to the calculation of the lot size, your account will be a non-starter regardless of your trading strategy. I hope that together we will puzzle out how to calculate your risks.

So now read carefully the main rules of money management which I define for the cryptocurrency market.

10% of the risk in one trade. That is, if you enter one trade, take no more than 10% of the risk. If you enter five trades a day, then, respectively, you should divide 10% of risk by 5. Do not make frequent transactions, trying to earn more. You will lose a lot of money by paying commission fees to the exchange.

30% of cash on the account is the minimum you should always have. You should never own only cryptocurrency or only fiat money. Even when the currency price falls, you should have some cryptocurrency assets. Respectively, when the market grows, you should have fewer assets in Bitcoin and more assets in Ether and altcoins. The larger capitalization of an asset, the less volatile it is. Thus, if you want to have smaller volatility in the fall, keep some of the money in Bitcoin. May it be 10-20% of all your funds, but you should not keep everything in cash. If you have Bitcoin, you can buy altcoins for it as they are often traded precisely for Bitcoin. At the same time, you should have at least 30% of cash, even when the market sees a super growth. You may need money for a new interesting ICO, a new movement with some kind of altcoin, *etc.*

The reason for entry should be equal to the reason for exit. Whenever you enter a trade, you should know when you will exit. You should have an exit plan for any situation. All your plans should be clearly spelled out on paper, and you should not retreat from them.

Diversification of marketplaces and assets. If you trade, you should do it on at least two top exchanges. Do not take risks even if the newly opened exchange seduces you with small fees or you are offered a bounty.

Do not make many trades and avoid gambling. Here you need to understand how often you open the terminal and monitor the quotes. I recommend that you look at the cryptocurrency movements not more frequently than once a day. Choose the time of day when you feel comfortable to do this. For example, you come home from work, pour a glass of red wine (or a cup of tea), and calmly sit down to see what's new in the cryptocurrency market. You do not need to follow the market every ten minutes convulsively. Such nervous behavior will do good neither for your health nor for your wallet.

Have your own opinion, and do not listen to the crowd. There is not a single person in the world who will tell you for sure how much Bitcoin will cost tomorrow or make other accurate forecasts. Therefore, if you hear very high-profile forecasts in favor of a particular coin, this person is probably biased in some way.

I once again remind you that nothing can be 100% certain. I suggest you always being reasonably sceptical. Nobody knows what will happen in the future. Remember this and see through the trick. If someone convinces you of something, ask yourself why this man proves his point so fiercely. You can listen to the opinion of other people, but you cannot invest being guided by their opinion.

Keep the stops in your head. As we said earlier, it makes no sense to use stop orders in the cryptocurrency market as they may not work or work badly. Therefore, always keep your stop (your exit point) in your head, or write it down in a notebook. For example, I bought the Ether for $250, and I will close the position at $200.

Risks. I am sure that you have clearly understood after reading this book that the cryptocurrency market is very risky. But you will not make such profits anywhere else. You cannot double or triple a sum in a day anywhere else.

Finally, I remind you that there are no guarantees in the cryptocurrency market. If a person gives you a guarantee, this person is a liar. Be ready for this as such situations are quite common in this market. The most cunning and bright-headed people (creators of ICOs, experienced traders, and so on) compete in the cryptocurrency market, and each of them pursues their own aims.

CHAPTER 15. INVESTOR MINDSET AND USEFUL TIPS

You probably agree that the psychological mindset behind any endeavor you want to succeed in is extremely important. Some people have the *right* mindset since childhood while others must continue to work to develop it.

Either way, I want to highlight a few key points you can use to make a profit in the cryptocurrency market or to at least avoid large losses.

Don't be greedy. Do not wait for the price to grow even higher so that you can profit even more. Your greed may result in you not only in missing out on profits but bearing a loss as well.

Be patient. If you buy a currency at a certain price and don't see strong up or down movements, do not fall into panic and sell it immediately. I know a lot of examples when the almost stable price of a cryptocurrency suddenly increased tenfold in a week. You may ask how not to meet a loss when the market tumbles. My answer is: do not change all your cryptocurrency for fiat! Many people make this mistake and deplore it later. Before selling everything, you'd better monitor thematic forums and user reviews on this coin.

Always take account of the market depth. If there are many sellers and few buyers, you can purchase some assets at a seemingly attractive price but fail to sell everything you have purchased.

Track the trading volume and cryptocurrency capitalization. You should do this to know whether there will be spikes. Thus, having analyzed the cryptocurrency price changes and having found certain regularities, you can create your own strategy and trading plan. You can start to refine it in real conditions, starting with small amounts allowed on almost all exchanges. If your strategy proves to be successful, you may gradually increase your investments.

Finally, I once again outline my short and effective tips. I hope they will help you become a successful cryptocurrency investor.

☐ Don't change strategy at every turn

☐ Take your profits

☐ Don't forget about your initial goals

☐ Don't make hasty decisions

☐ Automate the processes as much as possible.

CONCLUSION

At the very beginning of this book, I wrote that many people do not take cryptocurrency seriously because of the many myths surrounding the market. Let's summarize these myths now and finally dispel them.

First, a lot of people think that the most important and attractive thing in the cryptocurrency market is the potential for large profits and enter the market only for this reason. However, in my opinion, the profit is an absolutely secondary thing, which you can make in any business. The question is: what risks are you willing to take? Therefore, the most important topic in cryptocurrency is risk.

At the same time, it is said that the ratio of profits to risk is very distorted in the cryptocurrency market. Some people suppose that if you take a small risk in any other market and get a small profit from it, then here you get hundreds of times more profit when taking the same amount of risk. It is not true. The ratio of profits to risks is always approximately average. If it was really possible to get a large profit with low risks, Swiss banks would be emptied of all banknotes because everybody would have already invest in this business.

Many others also argue that it is too late to dive into the cryptocurrency market because it has become popular recently and all the profits have already been made and divided. I agree only with the fact it is going to be tough on the newcomers because of all the cryptocurrency hype but is this the reason to think that it's too late to get into the market and get your tidbit? If you constantly study, strive and try, you will also start to make good money. Hard work always pays off. It may take you more time and resources, but a slogger will definitely be rewarded with success.

Apart from the opinion that it is too late to enter the market now, some people say that only programmers or professional traders can work in this market. This is a myth. If you are a doctor, a teacher, or a middle-rank manager, you also can earn big dough in the cryptocurrency market. One can learn the necessary information and acquire the necessary skills no matter what their current skills. Nobody is born to be an artist, but you can become an artist by mastering different drawing techniques.

By and large, I do not know what conclusions you have drawn for yourself after reading my book. I only hope that you at least now know that cryptocurrency is a present-day reality, not a matter of the future. The sooner you understand this, the further you can tear yourself away from the sceptics who run behind the locomotive. Therefore, I strongly advise you to say "yes" to cryptocurrency.

If you read this book from front to back, I also wholeheartedly believe that I've helped you not just to sort through the cryptocurrency myths and basics but also helped you create a foundation for profitable long-term business in this area. A keystone has already been laid, but this is just the beginning. Now everything depends on you. Interpret the knowledge as it is possible to learn and receive something only by making attempts. Make first steps in the cryptocurrency market using the least risky strategies.

Now I want to address those who are sure that there is no need to make special efforts in this business. You may think that it will be enough for you to press the "start" button and money will fall from the sky. If you have already imagined yourself sitting by the ocean hoping that the cryptocurrency business will allow you to travel all the time without the need to work, you are very wrong. This business, like any other, requires that you get off the couch and put your strength, brains, and money to work to create a successful future.

The main thing required from you is to believe in yourself and in cryptocurrency. After all, it's those people who believed in Bitcoin a few years ago that now have a huge fortune and influence today. Therefore, believe in the dawn of the cryptocurrency economy!

And a few tips in conclusion.

First, you shouldn't take cryptocurrency too seriously. The cryptocurrency market is very volatile, and if you do not learn to be easy-going about risks and losses, you will drive yourself crazy. It is better to have an easy-going attitude to this business from the very beginning. Personally, I think it is impossible to survive in the cryptocurrency market without humor, and even a swear wordor two□. At the same time, do not treat this business as a casino or a lottery. In the cryptocurrency market, you need to have a sound strategy and a strong sense of purpose.

Second, you have to be cautious and sceptical of working in this market as we've already discussed at the very beginning of the book.

So, my dear reader, I cherish a hope that, having read this book, you will:

□ stop just following cryptocurrency news and start to make money on this market

□ sort out all the subtleties of this field for good and be able to identify scammers in the cryptocurrency market

□ choose the way of making profit in the cryptocurrency market that suits you best

□ invest your money skillfully and earn a good profit

□ find a source of semi-passive income and maybe even quit the job you are sick and tired of

☐ establish dozens of useful contacts with investors and cryptocurrency entrepreneurs;

☐ fulfil your dreams

Made in United States
Troutdale, OR
11/03/2026

56852704R00046